1 MONTH OF
FREE
READING

at

www.ForgottenBooks.com

By purchasing this book you are eligible for one month membership to ForgottenBooks.com, giving you unlimited access to our entire collection of over 1,000,000 titles via our web site and mobile apps.

To claim your free month visit:

www.forgottenbooks.com/free905512

ISBN 978-0-265-89214-5
PIBN 10905512

This book is a reproduction of an important historical work. Forgotten Books uses
state-of-the-art technology to digitally reconstruct the work, preserving the original format
whilst repairing imperfections present in the aged copy. In rare cases, an imperfection in
the original, such as a blemish or missing page, may be replicated in our edition. We do,
however, repair the vast majority of imperfections successfully; any imperfections that
remain are intentionally left to preserve the state of such historical works.

Journal

of

The New York Botanical Garden

VOLUME XXII, 1921

PUBLISHED BY THE AID OF THE

DAVID LYDIG FUND

BEQUEATHED BY CHARLES P. DALY

JOURNAL

OF

The New York Botanical Garden

EDITOR

R. S. WILLIAMS

Administrative Assistant

VOLUME XXII

WITH 13 PLATES

1921

PUBLISHED FOR THE GARDEN

AT 8 WEST KING STREET, LANCASTER, PA.
INTELLIGENCER PRINTING CO.

TABLE OF CONTENTS

JOURNAL

OF

The New York Botanical Garden

EDITOR

R. S. WILLIAMS
Administrative Assistant

CONTENTS

PUBLISHED FOR THE GARDEN

AT 8 WEST KING STREET, LANCASTER, PA.

INTELLIGENCER PRINTING COMPANY

JOURNAL

OF

The New York Botanical Garden

| VOL. XXII | January, 1921 | No. 253 |

REMINISCENCES OF ALVAN WENTWORTH CHAPMAN

In answer to the door-bell early one morning in December, 1917, I was brought face to face with a stranger standing on the porch of our house in Apalachicola. The visitor announced that he had been attracted by a plant of a genus foreign to this part of Florida, growing in our front yard. Would we give him some information concerning it? I started out by telling him the botanical name of the plant, that it was a cycad, and that it had relatives in the West Indies and in the Old World.

"I know all that myself," he answered patiently. "What I want to know is, where did you get it?" I opened my eyes a bit and really looked at him. "If you know *that* much you do know something about plants," I remarked. He nodded and announced himself as a member of the staff of The New York Botanical Garden.

He was then told the plant came from peninsular Florida, whence Doctor Chapman had brought it many years ago.

After having thus been introduced by *Zamia* I related some incidents of Doctor Chapman's life. Doctor Small was interested. He considered my idle memories of value; and he requested me to write some of these incidents and some of my experiences with Doctor Chapman, so that they might be given to others. Hence the following notes:[1]

[1] This document, prepared at the request of the undersigned, is a welcome addition to the too scant Chapman biographies and is the first written by one who was an intimate friend of the subject of the sketch, as well as the first to give the personal side of a man who was isolated from the other botanists of America. It is particularly fitting that it be published in our *Journal*,

I saw Doctor Chapman[1] first in April, 1887. We had come to Apalachicola on a visit, as my father was doing some Government work in the harbor, and he thought that this queer old abandoned city would interest my mother. So he brought his family and installed us all in the "Fuller House." That tavern is worthy of note in itself. It was a private residence before the Civil War, and was left, entirely furnished as it stood, by Madame Fuller upon her death, to "Uncle William" just prior to "Lee's Surrender." Now "Uncle William" was her husband's body servant, and during all those perilous days, first when her husband was at the front and then after his death, this negro stood at her side, ran all her affairs, and at times provided the very wherewithal of life. So when she came to the "Crossing," she left everything she possessed to this ex-slave, and now "Uncle William" ran the best hotel in town. Here all the "old timers" gathered, while the Yankees went to the hotel run by "po' white trash."

Doctor Chapman took his meals here. One day he saw my mother working around a syringa bush—one of the relics of a

as Doctor Chapman's main and original herbarium, the one on which the first edition of his "Flora of the Southern United States" was based, is now among the collections at The New York Botanical Garden. Among his books at the Garden is his "Catalogue of Southern Plants," a manuscript which was the basis of his "Flora."—JOHN K. SMALL.

[1] Alvan Wentworth Chapman, son of Paul and Ruth (Pomeroy) Chapman, was born September 28, 1809, at Southampton, Mass. He entered Amherst College in 1826, graduating (B. A.) in 1830; from 1831 to 1833 he was a teacher in a family on Whitemarsh Island, near Savannah, Ga.; in 1833 he was elected principal of the academy at Washington, Ga., and while there began the study of medicine, which he continued upon his removal to Quincy, Fla., in 1835, where he began to practice his profession the next year; in the summer of 1837 he moved to Marianna, but after a few months returned to "Rocky Comfort," near Quincy; in 1847 he settled at Apalachicola, where he practiced medicine for more than half a century. His interest in botany began before he left New England; the genus *Chapmania* was named in his honor as early as 1838; his "Flora of the southern United States," first published in 1860, and running through several editions, was for nearly fifty years the only manual of flowering plants of the southeastern states. He married in November, 1839, Mrs. Mary Ann Hancock, daughter of Benj. Simmons, of New Berne, N. C.; she died at Rome, Ga., in 1879. In 1886, the University of North Carolina conferred upon Chapman the degree of LL.D. He died at Apalachicola, April 6, 1899, in his ninetieth year.—JOHN HENDLEY BARNHART.

once beautiful garden—which was struggling for life against overwhelming odds. "The Doctor," as we soon learned to call him, came over and said, "Madam, I see that you love flowers."

From that day on an intimacy, then established between this grand old man and my family, held firm until his death in his ninetieth year. For in time business brought our family permanently to Apalachicola and we built a home here.

We built a home; and my mother planted a garden![1] Oh, what a delightful time she and "The Doctor" did have—experimenting with impossibilities! When a South American ship brought up clay from the "Rio de la Plata" as ballast, and my father had it spread over the garden, Doctor Chapman's interest grew apace. He watched each new "weed" that cropped up.

It was about this time that I began to study botany with him, and to accompany him on his long jaunts through the pine woods and deep into the titi swamps.[2]

I can see him just as he used to look as he walked on ahead carrying the japanned box of specimens; for he was one of the gentlemen of the old school, and sturdy a girl as I might be, my

---

[1] Among the plants early introduced into that garden and still thriving is the *Zamia* already referred to. Whether it came from the northern part of the peninsula or from the southern part of the State, will, perhaps, never be known. It is typical *Zamia floridana*, which grows more or less abundantly in pinelands from the latitude of Gainesville to the lower Florida Keys. It is not now generally known that Doctor Chapman once visited Key West. However, in a letter to Doctor Torrey, written from Apalachicola, 1847, he said, "I saw Dr. Blodgett at Key West four years ago . . . ." Thus, the plant in question may have come from the northern part of the peninsula, which region Doctor Chapman doubtless visited, or he may have brought it back with him from Key West. This plant was, very likely, native on Key West before its pinelands were devasted, or it may have been cultivated there from plants derived from the nearby Miami region, which is the center of its distribution as far as abundance is concerned. At any rate, the plant has long been grown at Apalachicola where it thrives, as is attested by specimens in the garden of Doctor Chapman's old house, on his grave, in the garden of the Kimball house, as well as in other front yards in the town.—J. K. S.

[2] Besides the pinelands, the titi swamp or titi bay is one of the most prominent plant associations in the flatwoods in the Apalachicola region. The titi, a shrub or tree, known botanically as *Cliftonia monophylla*, is the conspicuous plant feature. When in flower the myriads of snow-white flower-clusters eclipse all other vegetation.—J. K. S.

sex forbade me to carry anything. Doctor Chapman was a very handsome man. He stood over six feet. His hair was of that fine white texture that is likened to snow, but in reality has no similarity; it was the kind of white hair that age and clean thought give to a man. A brain had to do clear thinking to be thatched with such white hair. His eyes were clear and blue, and with all the enthusiasm and purpose of youth in their glance. He was even a bit of a dandy and always wore dark blue suits. He was insistent that he was a New Englander, as he was by birth; and his profile spoke of that hardy Puritan stock. He was very dignified in his movements—he walked slowly with head slightly forward, a habit acquired from searching the ground for its shy flowers. When he did not have that tin box, his hands were clasped behind his back. He was very deaf, but his sight was remarkably good, with the exception noted below.

In his youth Doctor Chapman must have been a hard man— one very stern of purpose. I remember his telling my mother stories of the war that confirm this idea. "I was a Union Man" he said; "and right there my wife and I disagreed. For she was a Secessionist, and when she would not change nor see my point of view, we agreed to part. She went to her home in Marianna. I stayed here. I never saw her during those four years. Yes, I heard from her once. She sent me four shirts. They were very good shirts, and I thanked her for them. That was all. When the war was over she returned. Did I miss her? Yes, some—but, my dear friend, those were very stirring years."

Yet he disliked to talk of those days, and many of the stories that I heard of that time came from a mutual friend, who was the son of one of the Doctor's patients and a satellite of his. This lad was so devoted and so faithful that he grew to be of great service, so that Doctor Chapman relied upon him in many ways. For just because he was a Union man his life was proscribed, and when the guerillas swept through the city on their too numerous raids they made for his drug store every time. But the Doctor's friends always warned him. Then, if the time was short, he would betake himself to Trinity Church and spend the night there in hiding or for such time as the raids lasted. "That is why you will still find cushions in my pew," he used to say. "If I must hide, I decided that I might as well be comfortable."

You see, he was the only surgeon in the city, and that fact saved his life; for even his enemies had a wholesome fear of the diseases that were rife during Dixie's tumultous four years of war, and Doctor Chapman was a clever physician as well. "But," he said, ''I would not have given a sixpence for my life many times during those four years."

Andersonville prison was situated near the Flint River. All the escaping prisoners that trusted to the river, and survived the horrors of the trip, eventually reached Apalachicola. One of the Doctor's own slaves would notify him that some of these wretched creatures were hiding in the marsh. Then Doctor Chapman would find his boy friend, who had a boat, and that lad would scurry off to get this craft. All the boats that the Doctor himself once owned had been destroyed; so this boy's old dugout would be prepared, and when the night was darkest, he and the Doctor would row the Yankee soldiers out to the blockading vessels.

Often the boy went alone, for the Doctor was watched too closely to risk the trip. But it was fine sport for the lad. "Never have I been so royally treated as I was on board those Yankee cruisers," he said. "And the worst of it was, I had to go back home and hold my tongue when my mother told father I was entirely too young to join our town guard."

But to return to our subject.

"I did not do much botanical work in those days," the Doctor would say; "I was too busy doctoring. Still, dear madam, there is a lot of pother in this doctoring; and I find that my deafness is not entirely detrimental, since if I can't hear people's groans they won't send for me. My flowers are far more satisfactory. That reminds me: If people did but know it, fresh air and sunshine, just what keeps the plants growing, will keep most of us well. Why, unless it was a case of easing a soul in or out of the world I did my best practice with hot baths and bread pills."

I asked him when he began to really botanize. He answered, "My love for all flowers began to develop soon after I came

South.[1] I think I first began to give up my medical practice and devote more attention to plants when I went up to Marianna the summer after Appomattox and visited my wife. I used to walk through the fields a good deal, and I began to note the humors and peculiarities of the growing things." He smiled. "I had had quite enough experience with human ills by that time."

I am not sure but I think it was several years after this time that he met Doctor Torrey in Florida, and they ranged through the hills in the region where the Flint and the Chattahoochee Rivers meet to form the Apalachicola River.[2] I know that the story of the discovery of the *Torreya taxifolia*[3] by his friend Mr. Croom was very dear to his heart; and he was always trying to make the tree grow in other places.

"I can't do it," he would exclaim. "That tree will grow around that locality and it won't grow elsewhere. Torrey tried to grow it North, but he never succeeded. Transplanted specimens sulked along in several places, but not many plants lived. And they make fence posts of it at Alum Bluff!"[4]

[1] Doctor Chapman's first letter to Doctor Torrey was dated "Gadsden Co., Fla., January 12, 1835" and mailed at Quincy the same day. This and four more letters before the end of the same year, attest his botanical knowledge and enthusiasm at that time. Torrey & Gray, Flora 1: 355. 1838, mention Doctor Chapman as "an accurate and indefatigable botanist."—J. K. S.

[2] Doctor Torrey made a pilgrimage to the Chattahoochee country in the winter of 1872, the year before his death—many years after the tree in question had been discovered and named in his honor. He never published an account of his visit and there is no evidence that he traveled as far south as Apalachicola.—J. K. S.

[3] Under the present rules of botanical nomenclature this tree is known as *Tumion taxifolium*.

[4] The tree, evidently, does not transplant and thrive readily. However, there are several successful instances on record: Two seedling plants were given to Doctor Torrey at New York by Mr. Croom the discoverer. One of these, sent to Newburgh, New York, was lost sight of; the other, planted in Central Park, New York, grew, and furnished sprigs of foliage which were worn by members at the formal organization of the Torrey Botanical Club (1867) and at Doctor Torrey's funeral (1873). This tree is, apparently not now living. Dr. Asa Gray records in the American Agriculturist for 1875 that, "I have been told that two seedlings of *Torreya* which Mr. Croom planted near his house at Quincy, Florida and which had become stately trees, have recently been demolished by the present proprietor; also that a tree of Mr. Croom's planting still flourishes in the grounds of the state-house at Talla-

He was kind and patient, and as a teacher was most interest-ing. He made many plans for me, which in my youthful egotism I rejected. I know now that I should have done well had I followed this friendly counsel. He wanted me to take up his work, so that there should be some one here to wear the mantle of his life research. When I failed him he closed in upon himself; for he had no children of his own.

Mrs. Chapman was a widow with one daughter when Doctor Chapman married her. This daughter married Doctor Chap-man's most intimate friend, and the two children of this marriage were treated as if they were his own granddaughters. When one of these young ladies visited Europe she had the entrée into the highest social circles. The second brother of the old German Emperor William I was an eager botanist, and he be-came a friend of this young lady. All court circles were opened to her. This prince became so enamored with Doctor Chap-man's ward that he proposed a morganatic union which, when the Doctor got wind of it, so incensed him that he cabled for the stepchild's immediate return to America.

My memories of those days when I studied with Doctor Chapman grow more and more disconnected, and I find myself putting down things which happened in the order of their per-sonal interest—not at all as they occurred. For instance: one day I went in to see him and he was at work upon some kind of fern. Botanists must excuse me for not being more technical. He was studying the spores of these plants, and all the work was done under a microscope. He looked up as I entered. "Come here, my child," he said, " and look at these."

I looked, and I grant you that I saw nothing interesting in those brown particles as the glass revealed them.

But he continued: "Behold the hand of the Almighty at its incredible task! The spores of these particular ferns are all coordinated in multiples of four. Never an exception. And if the All Seeing has carried his law of order to even these

hassee." "There are three or four specimens of Torreya on the Capitol grounds at Tallahassee and many in other parts of town," according to inform-ation recently received from Doctor R. M. Harper. Of course, Quincy and Tallahassee are not far from the general region for *Torreya*. A paper on *Torreya* and its distribution was published by Doctor Chapman in the Bot-anical Gazette **10**: 251–254. 1885.—J. K. S.

minute spores, has he not so governed the whole world that no man need fear?"

That was as near as he ever came to being "religious." In fact, he was called an "irreligious" man. That was because in his day men who did not underscore every letter of the Credo were called heathens.

"The pendulum had to swing back the other way," he would exclaim, "My father was a 'blue Presbyterian,' predestination, damnation, and all the rest made life, the Sabbaths especially, miserable to me. They pulled that old pendulum too far, and they made us what we are."

"If I have a creed," he said to me on another occasion, "it is summed up in these words:

'He prayeth best who loveth best
All things both great and small;
For the dear God who loveth us,
He made and loveth all.' "

On his return from a visit north, where he met Doctor Gray for the last time, he told us of a visit that the two friends made to Doctor Chapman's old home.

"We went all over the old place," he said. "It looked very much as it used to when I was a boy. They don't change much up there. When I walked out into the garden, there I saw the brass sugar kettle of the old times. When my eyes rested on that kettle I saw it just as it looked over sixty years ago when they put me to stir the boiling syrup—and I let it burn."

His face hardened as he remembered his injury.

"It was an unmerciful licking I got," he said, "and an unjust one. When I saw that kettle my gorge rose—I said out loud, 'Damn that kettle.' How Gray laughed when he understood!"

Doctor Chapman told me one evening of another incident of the same trip—one that was to remain very dear to his heart, for Doctor Gray died the season following. It seems that he and his friend, Gray, were particularly interested in violets, and Doctor Chapman said that there was a violet in New England that Gray did not have in his collection. Now, botanists will understand all the significance of such a declaration, for it became imperative for Doctor Chapman to prove his statement.

So he made a trip up to the country town where he had remembered finding this shy sister. It was late in the season, but he hoped to find one single specimen. He searched and searched, but could find nothing. He gave up and was returning to catch his train. The matter would have to lie over another year, but, just before he came to the station, right there in a corner of the fence, he found his violet—a perfect specimen in every botanical way.

"How did it get there?" he cried. "You need not tell me my eyes weren't directed to it. Just remember the multiples of four of those spores! Anyhow, I took off my hat and I said, 'I thank You!' " Now the violet is in Doctor Gray's herbarium.

Doctor Gray visited Doctor Chapman down here. It was before our acquaintance began; but I have heard him tell this story of his friend's visit:[1] "I had been writing Gray that I had found a new *Rhododendron*, and between you and me I really think it was this plant that brought him South. Anyway, we went after it. Fortunately, he came early enought to catch the pretty witch in bloom. I took him out to the reaches where the flower grew. It always looked white to me, but Gray said it 'blushed pink.' He looked at it several moments, then he knelt down and studied it closely. When he arose he came and offered me his hand. 'You are right,' he said. 'I never saw

[1] "I will not here give any account of a delightful ten days' episode, [at Apalachicola] . . . . Then followed a week and more at dead and dilapidated, but still charming, Apalachicola, where the post-office opens Monday evenings, when the steamboat arrives, and closes for a week the next morning, when she departs,—where the climate, thanks to the embracing Gulf, is as delicious in summer as it is bland in winter; where game, the best of fish, and the most luscious oysters are to be had almost for nothing, and blackberries come early in April when the oranges are gone; and where, far from the crowd and bustle of the world, with Bill Fuller for caterer, and his wife Adeline for cook, the choicest fare is to be enjoyed at the cheapest rate. Then there was the pleasure of renewing our acquaintance with Dr. Chapman, and botanizing with him over some of the ground which he has explored so long and so well; of gathering, under his guidance, the stately *Sarracenia Drummondii* in its native habitat, and, not least, acquiring from him fuller information respecting the localities where *Torreya* grows." Asa Gray in American Agriculturist 34: 266–267. 1875. Doctor Gray gives, in the paper from which the above paragraph is quoted, a full account of his visit to stations for *Torreya*. However, he never wrote, at least never published a further account of his experiences at Apalachicola.—J. K. S.

this species. I congratulate you on *Rhododendron Chapmanii.*' "

This incident brings me to a peculiarity of the Doctor's sight which was a great cross to him. He could not detect the color red in any of its shades. Blue was perfect, but a red rose for him was so dull that it looked like a dusky brown, and pinks were all white. This failing was the cause of many of our interviews, for he always came to my mother or to me to have his flowers described in shade and hue.

An active mind will surely keep itself as well acquainted as possible with all the literature that relates to its own metier; but it is an unusual one that will pick up a foreign tongue at the age of eighty years and in six months learn it so that the student can read books and letters sent to him. Gladstone, I believe, was one of the few men who has done a thing like this, but our doctor was even older than he. For when he was eighty he started to study German.

"I get so many letters and books from my German co-workers; and I want to read them myself," said he.

In six months he had acquired all the German he needed to carry on his voluminous correspondence with the botanists of that country; for Doctor Chapman was as well known abroad as he was at home.

He, of course, read Greek and Latin, and he was a lover of French. He had a fine old library, but being very generous with his books it is only the skeleton of it that went to Biltmore, North Carolina, and is now at Washington. Why, one day I was browsing through his books and I called out: "Oh, Doctor, let me take your Josephus home, I want to read it." He turned to me, puffing away on his pipe—for he was an inveterate smoker—"Take them all and keep them, my child, for I don't know of anyone else in this town who would be fool enough to want to read them!" I have the books today and I prize them. Still, Doctor Chapman was right. That was twenty years ago, and no one else has ever asked to borrow them!

There is one other story that he told us that is of interest to more than just his personal friends. He was a great friend of Doctor Gorrie,[1] perhaps the dearest, nearest friend that great

[1] John Gorrie was born at Charleston, South Carolina, October 3, 1803. He settled in Apalachicola as a physician in 1833, and resided there until his death, June 18, 1855. Artificial ice had been prepared in an experimental

scientist had, and his account of the discovery of manufactured ice holds enough interest to bear recording.

"One day Gorrie came to me back of the prescription counter, where I was making bread pills, I believe, and his eyes danced as they always did when he jibed at his own profession. He was sort of fussed up. He said: 'Come on over to my workroom—I believe I've stumbled on *it*.' Now, I knew that he was working hard trying to find a way to cool hospital wards in our tropical climate," Doctor explained to us—and we well understood how such a need would interest both these physicians.

"So," he continued, "I asked him, 'have you got it?' We crossed over to his home, and there he showed us the first piece of manufactured ice. He had stumbled on it indeed! 'You've got a great thing here,' I exclaimed. 'Yes,' he laughed, 'and I'm going to get those Yankee cotton-brokers. The ice-ship is late—no ice for the dinner they are giving the English brokers. We'll wager that there will be ice for that dinner—and I'll make the ice!' He did it. We won the wager. And we had the ice! It was a great dinner we had that night."

"How much did Doctor Gorrie make from his invention?" I asked. Doctor Chapman shook his head.

"Relatively nothing. He was no business man, was Gorrie. If he had been he never would have invented artificial ice!"

When I survey my acquaintance with Doctor Chapman, I am impressed with the sweetness of it. Never a thing to repel you; yet he has been called a very stern and haughty man. He was not. Only, if you interest yourself with plants, flowers, and all the teeming life of the world, you have little time for the pettiness of people. He lived, and gave himself to his profession and to a few friends; to the rest of the town his deafness served as an excuse for an isolation that he welcomed.

WINIFRED KIMBALL

---

way for more than a century, but he was the first to invent and patent an ice-machine for practical use. Recognition of Gorrie as one of the great benefactors of the human race has been slow. Even now, when there is a monument to his memory at Apalachicola and a statue in the Capitol at Washington, one may search through the long article upon "refrigeration" in the latest edition of the Encyclopaedia Britannica without finding any mention of Gorrie's name.—JOHN HENDLEY BARNHART.

WILD FLOWERS OF NEW YORK

The New York State Museum has recently completed the publication of its Memoir No. 15, containing descriptions and illustrations of selected wild flowers inhabiting the state, written by Dr. Homer D. House, State Botanist. This Memoir is issued in two parts, of about equal size, containing collectively 264 plates printed in color, reproduced from photographs taken by Dr. House in various parts of the state, and there are a number of text figures as well. The descriptions are drawn so that any person of ordinary scientific education can understand them without effort. The distribution and habitat of the plants illustrated are given. There is a complete list of illustrations and a comprehensive index.

The excellent photographs taken by Dr. House are reproduced for the most part with great accuracy and the color reproduction is very true to nature. The work is a noble contribution to the natural history of the state and cannot fail to stimulate observation on its wild flowers and their relation to soil and climate, and also to increase interest in the preservation of many of the rarer and more beautiful species which locally are in danger of extermination by indiscriminate picking.

It is with unusual pleasure that we are adding these two elegant volumes to the library of the Garden, and express our appreciation of this work to Dr. House, who was a student of the New York Botanical Garden in 1903–1904, a museum aid here in 1907–1908, and has been State Botanist of New York since 1913. At the same time we would congratulate Dr. John M. Clarke, Director of the State Museum, on the successful production of this important work, so largely due to his interest and encouragement.

WILLIAM HARRIS

Mr. William Harris, Assistant Director of Agriculture and Government Botanist of the island Jamaica, British West Indies, for many years a highly valued correspondent of the New York Botanical Garden, and an efficient cooperator in all field work in Botany carried out in Jamaica by students from England and the United States, died at Kansas City, Missouri, on October 11, 1920. He had been ill for several months, and came north early in September for medical treatment, but his malady proved to be incurable; he was born at Enniskillen, Ireland, November 15, 1860, and was thus about sixty years old and had been in the service of the Jamaica Department of Agriculture as Superintendent of Public Gardens and Plantations and in related branches of that government for nearly forty years; since arriving in Jamaica at about the age of twenty, after about two years' training as a gardener at Kew, he had never been away from there until his trip to the United States to be with his son in Kansas City this last autumn. All his official duties were carried out with devotion and fidelity, and he was a very successful and esteemed administrator of the several botanical gardens and establishments of the colony, Hope, Castleton, Cinchona and Bath.

Harris was a born naturalist; he early became interested in the native vegetation of Jamaica and studied it intensively up to the time his fatal illness incapacitated him, serving practically as insular botanist in addition to his official gardening and agricultural duties; these took him widely over the island, in addition to visits to the gardens under his charge, and he knew the hills, mountains, savannas and coasts as well, doubtless, as they have ever been known by any one.

The care of the herbarium at Hope Gardens came under his charge, and he soon began collecting specimens to increase it, with such diligence that his collection has now become one of the most important .repositories of botanical information in tropical America; his botanical collecting was begun as early as 1891, when, assuming an arbitrary number 5000 to represent the approximate number of specimens in the Jamaica herbarium, he continually increased it up to 12909, preserving, as a rule, several duplicate specimens of each number, which found their way to museums in Europe and America, where they are highly valued; since 1904 his collecting has been in cooperation with the New York Botanical Garden, and his duplicate specimens have been mostly distributed from here to other institutions.

My personal acquaintance with Mr. Harris dates from the autumn of 1906, when I visited Jamaica to prosecute botanical investigations,[1] at which time he accompanied our party during a study of the rough Cockpit Country and elsewhere; we were with him again in the late summer of 1907, exploring the Santa Cruz Mountain Range and Westmoreland;[2] again in the spring of 1908, when we studied most of the coast-line and parts adjacent, using a schooner,[3] and also in the late summer of that year, studying especially in the Blue Mountains and in the parishes of Manchester and St. Thomas.[4] My last field work with him was in the spring of 1909, exploring the difficult John Crow Mountain Range,[5] and I have not seen him since that memorable expedition; our correspondence has been continuous however, and is, collectively, voluminous.

He was one of the most enjoyable scientific companions I have ever known, whether in the field or in the laboratory, always cheerful, active and original, and he would always find means of obtaining plants we wanted to see for one reason or another. His contribution to knowledge of the vegetation of Jamaica is highly important as a scientific achievement and his name will forever be associated with the subject. He is commemorated by a number of species, new to Science, named in his honor, by the Cactus genus *Harrisia* and by the Orchid genus *Harrisiella*.

N. L. BRITTON

[1] Journ. N. Y. Bot. Gard. **7**: 245–250.
[2] Journ. N. Y. Bot. Gard. **8**: 229–236.
[3] Journ. N. Y. Bot. Gard. **9**: 81–90.
[4] Journ. N. Y. Bot. Gard. **9**: 163–172.
[5] Journ. N. Y. Bot. Gard. **10**: 99–102.

BEQUEST OF MARY J. KINGSLAND

A bequest of five thousand dollars to The New York Botanical Garden by Mrs. Mary J. Kingsland, was received from the executors of her estate in the summer of 1920, and on recommendation by the Endowment Committee this gift has been appropriated by the Board of Managers for the construction of the new Horticultural Garden entrance on the Southern Boulevard, and for the boundary fence adjoining this entrance. After approval of plans by the Municipal Art Commission, work was begun upon these structures in the autumn and by the close of the year they were essentially completed, with the exception of the iron railings needed, which it is hoped to obtain, at reasonable cost, during the winter or early spring.

This entrance to the Horticultural Garden is for pedestrians only; it consists of two piers built of cut cast stone; on one of these a suitable bronze tablet will be placed recording Mrs. Kingsland's gift. The boundary fence follows the design of those previously built on the Fordham University boundary and along the Bronx Boulevard, and will add about four hundred feet to the permanent fencing of the reservation.

CONFERENCE NOTES FOR NOVEMBER AND DECEMBER

The November conference of the Scientific Staff and Registered Students of the Garden was held in the laboratory of the museum building on Wednesday, November 3, 1920, at 3:30 P. M.

The program was as follows: "Intersexes in grapes" by Dr. A. B. Stout and "The genus *Veronica*" by Dr. F. W. Pennell.

Dr. Stout reported on the types of flowers in grapes with special reference to fruit development. The report was based on studies that are being made at the State Agricultural Experiment Station at Geneva, N. Y., in cooperation with the Horticultural Department of the Station. Here are grown many named varieties of European and American grapes which afford

material for a study of flower types in varieties whose perform-
ance in fruit production is well known. There are also several
thousand seedlings blooming and fruiting that constitute 1st,
2nd, and 3rd generations of progenies from known parentage.
These seedlings afford most excellent material for studies of
variation in flowers and in fruit production and of the inherit-
ance of sex in grapes.

The general survey of this material revealed at least one new
type of flower, a wide range of variation in the length of stamens
among flowers classed as staminate and as hermaphrodites, and
various intermediates between the typically upright and the
reflexed types. In breeding a plant exhibiting such variations
in sex as are seen in the grape it becomes necessary to determine
as fully as possible how various types of fruit may be produced.
The breeder is concerned with understanding how the production of desirable types of flowers may be regulated, controlled,
or influenced by breeding and by the selection of parentage.

This problem is especially being considered in respect to the
production of seedless sorts. The studies indicate clearly that
these are characterized by strong maleness and weak femaleness. The results of the breeding work already obtained at
the Experiment Station indicate clearly that the use of seedless
and near-seedless plants as male parents in crosses with varieties that are strongly female (perfect and imperfect hermaphrodites) gives progeny that are strongly female and seedproduc-
ing. The first generation of the hybrid offspring of many
crosses of standard seeded varieties with Hubbard Seedless
have all been strongly pistillate, yielding seeded fruits. Weak
femaleness (seen in seedless fruits) is in this case dominated or
swamped by the strong femaleness of the seed parent. However, the seeded character of the individuals of the first generation is no index of the variation in intersexes that may appear
in later generations and the segregation of at least some plants
bearing seedless fruits may be expected. The use of other
seedless sorts in such crosses may, however, give different results.

The most effective course in breeding for the development of
seedless sorts is suggested by the conditions of intersexualism.
Most individuals and varieties producing seedless or near-
seedless fruits are strongly staminate. The former can be used

as male parents on the latter which do produce a few viable seeds. Plants strongly male can be crossed with plants strongly male but weakly female and also the self-fertilized progeny of the latter may be obtained. In this way families weak in femaleness may undoubtedly be obtained in which a considerable number of individuals produce seedless fruits.

A more detailed discussion of intersexualism in the grape, with descriptions of the various types of flowers and with several plates of illustration will be published in the near future.

Dr. Pennell reviewed his recent study of the American species of *Veronica* and its near allies of the Scrophulariaceae. He discussed mainly the trend of evolution within the group.

One group, long held as a section, must be accorded generic rank. *Hebe* is a polymorphous genus of shrubs and even small trees, of New Zealand and southern South America, distinguished by having flowers localized in axillary racemes, and a capsule the septicidal dehiscence of which precludes its having been developed from that of the other "Veronicas."

Veronicastrum (*Leptandra*) is also considered a distinct small genus of eastern Asia and eastern North America. It is certainly an ancient type and the tubular corolla represents that from which the rotate corolla of the true Veronicas have developed. It is an erect, tall herb.

Veronica itself consists of low herbs, primitively perennial, erect, with opposite leaves, terminal racemes, five sepals and capsule not notched though even in the most primitive types it is already loculicidal. Tendencies from this are to plants annual, depressed and spreading, the leaves mostly alternate, the flowers localized in axillary racemes, the sepals four and the capsule strongly notched and bilobed. The chief cleavage is between the plants with flowers in generalized terminal racemes, Veronicella, and those with flowers in specialized axillary racemes, Euveronica or "true Veronica."

Dr. T. Harvey Johnston of the University of Queensland, Australia was present at the Conference and spoke very interestingly and instructively on his mission to the United States in connection with the work of the Prickly Pear Travelling Commission. Several species of cacti, commonly called pricklypears, have escaped from cultivation in Australia, and have now become serious weed pests. They spread rapidly by both seed and vegetative propagation and quickly occupy tillable land and even land under cultivation. Thousands of acres of agricultural and grazing lands are already covered with a growth of these cacti. The Government of Australia has established a Commission to consider means of control and extermination of this pest. The Commission has now been actively at work since 1914 and has decided that the most promising means of control is the destruction by natural enemies, especially by insects and fungi. A considerable number of natural parasites are known to be very destructive only to cacti and the Commission is now searching in many parts of the world for such parasites as are suitable for introduction into Australia.

The December conference of the Scientific Staff and Registered Students of the Garden was held in the laboratory of the museum building, on December 1, 1920, at 3:30 P. M.

The program was as follows: "*Siphocampylus* and *Centropogon* in South America" by Dr. H. A. Gleason. "The Genus *Diphysa*" by Dr. P. A. Rydberg.

Dr. H. A. Gleason reported on "*Siphocampylus* and *Centropogon* in South America." These two genera of Lobeliaceae are strictly American, with their center of distribution in northwestern South America, and extend south throughout tropical and subtropical South America. Northward they are distributed in smaller numbers as far as Mexico and through the West Indies. The species, about 400 in number, are beautifully characterized, mostly of local distribution, and probably represent recent and rapid evolutionary development. The two genera are distinguished by capsular and baccate fruits, respectively, but the two groups thus separated show such extraordinary parallelism that it seems probable that this character is not of generic value. A tentative separation has been made on the basis of corolla structure, by which four or five groups may be distinguished. The speaker's studies do not as yet warrant a decision as to the generic value of these characters. Most of the species are shrubs or vines with large, showy, brilliantly colored flowers. For this reason they early attracted the attention of botanists and horticulturists, and several species have been introduced into cultivation. The group as a whole appears to offer exceptional advantages for the correlation of geographic distribution with specific differentiation.

Dr. Rydberg reported on his work on the genus *Diphysa*, illustrating his remarks with specimens and sketches on the blackboard. The genus belongs to the family Fabaceae and to the tribe Galegeae and has usually been placed near *Robinia*. Dr. Rydberg thought that it was rather more closely related to a small group of genera clustered around *Sesbania*. This group consist of *Sesbania*, *Agati*, *Daubentonia* and *Glottidium*. In this group as well as in *Diphysa*, the structure of the corolla and the calyx proper is practically the same as in *Robinia*, but the fruits are quite different. In *Diphysa* as well as in *Sesbania* and its relatives, the calyx is subtended by two bractlets which are wanting in *Robinia* and most genera of Galegeae. The fruit in *Diphysa* resembles that of *Sesbania* and its allies in this respect that the seeds are separated by a more or less distinct false partition, and is especially like that of *Glottidium* in that the pericarp separates into two layers of which the endocarp closely invests the seeds and the exocarp is thin and loose. In *Diphysa* the latter is strongly inflated and forms two elongate bladders, one on each side. *Diphysa* differs in another respect from both the Robinianna and the Sesbanianeae in the presence of a well developed obconic hypanthium below the calyx. *Diphysa* has, as *Robinia*, odd-pinate leaves while in the Sesbanianeae the leaves are abruptly pinnate.

The type of *Diphysa* is *D. carthageniensis* described from Cartagena, Colombia. This is the most southern of the species. The genus ranges from Colombia to Arizona and has its best development in Southern Mexico. The most northern species was described by Dr. Gray as *Daubentonia Thurberi*. Its fruit has been unknown until recently. As the plant has odd-pinnate leaves and does not resemble much the known species of *Daubentonia* but more that of the Old World genus *Ormocarpum*, Hemsley transferred it to that genus. Dr. Otto Kuntze transferred it in one and the same publication both to *Emerus* and *Solueus* because he thought that the former name should replace *Sesbania* (including *Daubentonia*) and the latter *Ormocarpum*. The recently collected fruiting specimens show that the species is a true *Diphysa*, related to *D. racemosa*, but with narrower pod than the other known species.

A. B. STOUT
Secretary of the Conference

NOTES, NEWS AND COMMENT

Under the benches in one of the orchid houses at Conservatory Range No. 2 is to be found a most interesting plant, *Dorstenia contrajerva*. This is a member of the mulberry family, and bears its flowers and fruit in a curious receptacle formed of the upper end of a long green scape arising from the crown. The structure is like a fig split open. A native of Tropical America, this plant seeds itself readily in our warm houses.

The second volume of "The Cactaceae," by N. L. Britton and J. N. Rose has recently come from the press. The first volume included two tribes, the Pereskieae and Opuntieae. This one embraces two more, the Cereanae of 38 genera, and Hylocereanae of 9 genera. It is a quarto of 239 pages with 40 plates and 305 text-figures. Published by the Carnegie Institution of Washington.

The Women's Auxiliary Board and the Board of Managers of the New York Botanical Garden will entertain the members of the Garden at the Waldorf-Astoria on the evening of February 8. The program will include short talks on various activities now in progress at the Garden.

Mr. Chas. C. Deam, State Forester of Indiana, widely known from his long and detailed studies of the Indiana flora, was a recent visitor at the Garden.

Mr. William W. Diehl of the United States Department of Agriculture, Washington, D. C., recently spent several days in mycological work at the Garden.

Professor C. R. Orton of Pennsylvania State College spent a few days in December at the Garden in continuation of his studies of certain fungi parasitic on grasses.

Dr. P. J. Anderson, of the Agricultural College at Amherst, Massachusetts, spent a few days at the Garden looking over the collections of cup-fungi preparatory to a monograph of the genus *Helvella* for his own state.

In the latter part of November, Dr. P. A. Rydberg made a visit to Washington to study the collections of Fabaceae, Carduaceae, and Solanaceae in the National Herbarium. This was done in connection with forthcoming monographs in the North American Flora. The trip lasted 10 days.

Dr. A. B. Stout, of the Garden Staff, served as a member of the Committee of Judges of the Mayor's Committee of Women in judging the exhibits of paper white narcissus grown by children of the public schools of Manhattan. About 8,000 bulbs were distributed under the immediate direction of the Americanization Committee and grown under the supervision of the teachers. The entire exhibit, which was a most creditable one making a beautiful display of bloom, was assembled at P. S. 27 on December 9, for the general inspection by the public and that evening prizes were awarded and appropriate exercises held under the auspices of the Mayor's Committee of Women.

Meteorology for November.—The total precipitation for the month was 3.00 inches. The maximum temperatures recorded for each week were as follows: *71° on the 1st*, 62° on the 9th, 55° on the 20th, and 48° on the 23rd. The minimum temperatures were: 38° on the 4th and 6th, *23° on the 13th*, 34° on the 19th and 30° on the 29th. There were traces of snow on the 15th and 25th. First ice of the autumn formed across the middle lake on the 14th and 15th.

Meteorology for December.—The total precipitation for the month was 4.73 inches, of which 0.20 inches (2 inches snow measurement) fell as snow. The maximum temperature recorded for each week were as follows: 56° on the 5th, 51° on the 12th, *58° on the 14th*, and 54° on the 23rd. The minimum temperatures were 29 on° the 3rd, 24° on the 8th, 21° on the 20th, and *14° on the 26th.*

Meteorology for the year 1920.—The total precipitation at the New York Botanical Garden for the year was 46.61½ inches. This was distributed by months as follows: January, 2.59½ inches (including 7 and ¼ inches snow measurement); February, 4.17 inches(including 22 inches snow measurement); March, 3.16 inches (including 4 inches snow measurement); April, 4.23 inches; May, 3.13 inches; June, 4.04 inches; July, 3.68 inches; August, 7.49 inches; September, 5.01 inches; October, 1.38 inches; November, 3.00 inches, including traces of snow; December, 4.73 inches (including 2 inches snow measurement). The total fall of snow for the year was 35 and ¼ inches.

The maximum temperature for the year was 93° on the 11th and the 15th of June. The minimum was —4° on the 1st of February. The first freezing temperature of the autumn was in the morning of the 13th of November when 23° was recorded. The report of 19° on Sep. 20 printed in the November number of the Journal is a typographical error; the record for that day was 39°. The latest freezing temperature in spring was on the morning of April 10th when the temperature of 24° was reached.

ACCESSIONS

PLANTS AND SEEDS

2 plants for conservatories. (Given by Mr. John Sommer.)
1 plant of *Heliconia Edwardus Rex*. (Given by Mr. W. B. Thompson.)
2 plants of *Acacia*. (Given by Mrs. F. A. Constable.)
1 plant of *Woodwardia*. (Given by Mr. H. W. Becker.)
6 plants of *Iris*. (Given by Thos. Meehan & Sons.)
4 plants of *Iris*. (Given by Mr. F. H. Horsford.)
2 plants of Everblooming Raspberry La France. (Purchased.)
18 grape plants. (Purchased.)
1 plant from Trinidad. (Collected by Mrs. N. L. Britton.)
2 plants of *Mamillaria prolifera*. (By exchange with Bro. Hioram.)
1 plant of *Mamillaria macromeris*. (By exchange with U. S. Nat. Museum, through Dr. J. N. Rose.)
214 plants for conservatories and nurseries. (By exchange with Bureau of Plant Industry.)
4 plants of Cacti. (By exchange with U. S. Nat. Museum, through Dr. J. N. Rose.)
9 plants of *Castanea mollissima*. (By exchange with Bureau of Plant Industry.)
48 cuttings of *Populus tremuloides vancouveriana*. (Given by Mr. G. W. Chance.)
3 cuttings of *Saccharum officinarum*. (Given by Mr. G. W. Chance.)
2 plants of *Sequoia washingtoniana*. (Given by Miss Mary N. Welleck.)
7 plants from Florida. (Given by Mr. F. F. Hunt.)
4 plants of *Iris*. (Given by Elm City Nursery.)
3 plants of *Caltha palustris*. (Given by Miss M. E. Eaton.)
2 plants of *Chrysanthemum arcticum*. (Given by Mr. W. J. Matheson.)
34 plants from Florida. (Given by Mr. J. J. Soar.)
1 plant of *Trillium cernuum*. (Given by Mrs. L. M. Keeler.)
3 plants of Box Berberis. (Given by Elm City Nursery.)
12 plants for conservatories. (Given by Mrs. F. A. Constable.)
3 plants of *Dracaena*. (Given by Mr. H. W. Becker.)
2 plants for nurseries. (Given by Mr. W. W. Ashe.)
1 plant of *Helleborus viridis*. (Given by Mrs. Tysilio Thomas).
10 plants of *Iris cristata*. (Given by Mrs. O. A. Runyon.)
58 plants from Florida. (Collected by Dr. J. K. Small.)
47 plants from Trinidad. (Collected by Dr. N. L. Britton.)
1 plant of *Cocos nucifera*. (By exchange with Barnard College.)

28 plants of Cacti. (By exchange with U. S. Nat. Museum, through Dr. J. N. Rose.)

1 plant of *Robinia grandiflora*. (By exchange with U. S. Dept. Agric. Forest Service, through Mr. W. W. Ashe).

236 plants for conservatories. (Given by Mrs. H. A. Arnold.)

5 rose plants. (Given by Mr. J. P. Sorenson.)

5 plants for conservatories. (Given by Mr. John Sommer.)

72 plants for pinetum. (Purchased.)

3 plants of *Opuntia*. (By exchange with Mr. J. P. Otis.)

59 plants of *Penstemon*. (Collected by Dr. F. W. Pennell.)

1 plant of *Bryophyllum pinnatum*. (By exchange with Dr. Walter Mendelson.)

11 plants of Cacti. (By exchange with U. S. Nat. Museum, through Dr. J. N. Rose.)

1 plant of *Tripsacum laxum*. (By exchange with Univ. of Indiana.)

196 plants derived from seeds from various sources.

203 Dahlia roots, 76 varieties. (Given by Dr. M. A. Howe.)

88 Dahlia roots, 81 varieties. (Given by Mr. W. J. Matheson.)

62 Dahlia roots, 31 varieties. (Given by Dahliadel Nurseries.)

32 Dahlia roots, 31 varieties. (Given by Mr. Alfred E. Doty.)

47 Dahlia roots and plants, 24 varieties. (Given by Mills & Co.)

24 Dahlia roots and plants, 24 varieties. (Given by Mr. C. Louis Alling.)

30 Dahlia roots and plants, 19 varieties. (Given by Miss Emily Slocombe.

27 Dahlia roots and plants, 19 varieties. (Given by Mr. J. J. Broomall.)

22 Dahlia roots, 22 varieties. (Given by Mr. Geo. L. Stillman.)

18 Dahlia roots, 10 varieties. (Given by Mr. Richard Vincent, Jr.)

19 Dahlia roots, 19 varieties. (Given by Mr. J. K. Alexander.)

25 Dahlia roots, 11 varieties. (By exchange with Mr. Edwin Marquand.)

23 Dahlia roots, 19 varieties. (Given by Mr. H. McWhirter.)

14 Dahlia roots, 14 varieties. (Given by Mr. F. C. Burns.)

11 Dahlia roots, 6 varieties. (Given by Mrs. Chas. H. Stout.)

15 Dahlia roots, 5 varieties. (Given by Miss Annie Lorenz.)

7 Dahlia roots, 7 varieties. (By exchange with Mr. L. H. Bubois.)

8 Dahlia roots, 7 varieties. (Given by N. Harold Cottam & Son.)

14 Dahlia roots, 14 varieties. (By exchange with Dr. W. A. Orton.)

3 Dahlia roots, 3 varieties. (By exchange with Garden Club of Ridgwood, N. J.)

3 Dahlia roots, 3 varieties. (By exchange with Mr. F. P. Quimby.)

28 Dahlia roots, 19 varieties. (By exchange with Mr. Joseph Smith.)

18 Dahlia roots, 3 varieties. (Given by Frank D. Pelicano & Co.)

2 Dahlia roots, 2 varieties. (By exchange with Mr. C. Frey.)

6 Dahlia roots, 6 varieties. (By exchange with Mr. Robert B. Goeller.)

1 Dahlia root. (Given by Mr. Joseph Pfluger.)

3 Dahlia roots, 3 varieties. (Purchased.)

1 Dahlia root. (By exchange with Miss Rosalie Weikert.)

1 plant of Begonia. (Given by S. C. Templin & Son.) ·

112 plants of *Iris*. (Given by Miss Grace Sturtevant.)

303 plants of *Iris*. (Given by Mr. John C. Wister.)

1 plant of *Iris*. (Given by Mr. F. H. Presby.)

1 plant of *Portulaca pilosa*. (Collected by Mr. Kimball.)

19 plants of Cacti for conservatories. (By exchange with U. S. Nat Museum, through Dr. J. N. Rose.)

5 bulbs of *Validallium tricoccum*. (Given by Dr. H. H. Rusby.)

56 plants derived from seeds from various sources.

26 plants of *Iris*. (Given by Mrs. C. S. McKinney.)

1 plant of *Rumex alpinus*. (Given by Miss E. M. Kittredge.)

3 plants of *Iris*. (Given by Mrs. O. A. Runyon.)

3 plants of *Persea*. (Given by Mr. D. S. Dark.)

2 plants of *Talinum Mengesii*. (Given by Mr. Bede Knapke.)

59 plants of *Iris*. (Given by Miss Grace Sturtevant.)

1 plant of *Iris*. (Given by Mrs. E. M. Sanford, through Mrs. C. S. Mc-Kinney.)

14 plants for nurseries. (Collected by Dr. J. K. Small.)

4 plants of *Acer pennsylvanicum*. (Collected by Dr. E. B. Southwick.)

10 plants of Cacti. (Collected by Dr. J. K. Small.)

1 plant of *Commelina*. (Collected by Dr. F. W. Pennell.)

2 plants of *Opuntia*. (Collected by Mr. R. L. Barney.)

2 plants of *Opuntia*. (Collected by Dr. J. K. Small.)

3 plants of Cacti. (By exchange with U. S. Nat. Museum, through Dr J. N. Rose.)

6 orchid plants. (By exchange with U. S. Nat. Museum, through Mr. Wm. R. Maxon.)

1 plant of *Cereus*. (Given by Mr. E. W. Poole.)

1 plant of *Iris*. (Given by the Crawford Garden.)

39 plants of *Iris*. (Given by Mr. C. H. Hall.)

6 plants of *Iris*. (Given by Mr. J. T. Lovett.)

57 plants of *Iris*. (Given by Mr. Willis E. Fryer.)

8 plants of *Iris*. (Given by Mrs. Ella F. Rockwell.)

15 plants of *Iris*. (Given by Peterson Nursery.)

2 plants of *Iris*. (Given by Mr. E. M. Buechly.)

43 plants of *Iris*. (Given by Mr. T. M. Fendall.)

16 plants of *Iris*. (Given by Mrs. Azno Fellows.)

30 plants of *Iris*. (Given by Mrs. C. S. McKinney.)

8 plants of *Iris*. (Given by Mr. H. P. Sass.)

12 plants of *Iris*. (Given by W. J. Engle & Son.)

160 plants derived from seed from various sources.

3 plants of *Iris*. (Given by Mr. Wm. S. Johnson.)

30 plants of *Iris*. (Given by Mrs. W. S. Rait.)

4 plants from Pennsylvania. (Collected by Dr. F. W. Pennell.)

1 plant of *Lacinaria scariosa*. (Collected by Mr. Percy Wilson.)

5 plants from Pennsylvania. (Collected by Dr. J. K. Small.)

8 plants of Cacti for conservatories. (By exchange with U. S. Nat. Museum through Dr. J. N. Rose.)

5 bulbs for conservatories. (By exchange with U. S. Nat. Museum, through Dr. J. N. Rose.)

68 plants for conservatories. (By exchange with Prospect Park.)

Provisions for
Benefactors, Patrons, Fellows, Fellowship Members, Sustaining Members, Annual Members and Life Members

---

1. Benefactors

The contribution of $25,000.00 or more to the funds of the Garden by gift or by bequest shall entitle the contributor to be a benefactor of the Garden.

2. Patrons

The contribution of $5000.00 or more to the funds of the Garden by gift or by bequest shall entitle the contributor to be a patron of the Garden.

3. Fellows for Life

The contribution of $1000.00 or more to the funds of the Garden at any one time shall entitle the contributor to be a fellow for life of the Garden.

4. Fellowship Members

Fellowship members pay $100.00 or more annually and become fellows for life when their payments aggregate $1000.00.

5. Sustaining Members

Sustaining members pay from $25.00 to $100.00 annually and become fellows for life when their payments aggregate $1000.00.

6. Annual Members

Annual members pay an annual fee of $10.00.

All members are entitled to the following privileges:

1. Tickets to all lectures given under the auspices of the Board of Managers.
2. Invitations to all exhibitions given under the auspices of the Board of Managers.
3. A copy of all handbooks published by the Garden.
4. A copy of all annual reports and Bulletins.
5. A copy of the monthly Journal.
6. Privileges of the Board Room.

7. Life Members

Annual members may become Life Members by the payment of a fee of $250.00.

Information

Members are invited to ask any questions they desire to have answered on botanical or horticultural subjects. Docents will accompany any members through the grounds and buildings any week day, leaving Museum Building at 3 o'clock.

Form of Bequest

I hereby bequeath to the New York Botanical Garden incorporated under the Laws of New York, Chapter 285 of 1891, the sum of..........

Vol. XXII   February, 1921   No. 254

JOURNAL

OF

The New York Botanical Garden

EDITOR

R. S. WILLIAMS

Administrative Assistant

CONTENTS

PUBLISHED FOR THE GARDEN

AT 8 WEST KING STREET, LANCASTER, PA.

INTELLIGENCER PRINTING COMPANY

In shore hammock on Terra Ceia Island. *Harrisia aboriginum* in foreground; hammock-jungle in background. The cactus plants are usually fifteen to twenty feet tall, and commonly much branched. They grow in a soil composed of shells piled up by the aborigines and sand blown in from the bay. In addition to these ingredients, in some places the fossil remains of the aborigines are abundant. The cactus is particularly interesting, as it is the only one this side of the Gulf Stream with brown-haired flowers and yellow fruits.

JOURNAL

OF

The New York Botanical Garden

| VOL. XXII | February, 1921 | No. 254 |

OLD TRAILS AND NEW DISCOVERIES

A RECORD OF EXPLORATION IN FLORIDA IN THE SPRING OF 1919

With Plates 253 and 254

Exploration of new and little-known parts of Florida which had so greatly enriched our knowledge of the *Opuntiae* thereabouts in recent years, convinced the writer that there was also much more to be learned in that State than was known prior to 1918, concerning the cacti of the *Cereus* group. His attention was directed especially to the genus *Harrisia*, for there was evidence of its occurrence in many parts of both the southern half of the Florida peninsula and on the keys. As this evidence had a timely importance, moreover, in view of the monograph of the genus *Harrisia* contemplated by Doctors Britton and Rose, it was deemed worth following up without delay. We therefore organized a cactus-hunt. The Board of Scientific Directors of the Garden approved. Mr. Charles Deering, to whom we are already so deeply indebted, once more put launch and automobile at our disposal for the field-work. And it is thus that we are able to make the following report on botanical exploration of Florida in the early months of 1919.

We were enroute for Miami, Florida, the last week in April.

The season in the North was backward; but, even so, the shrubs and trees had begun to don their green, either singly or en masse; various water-plants, both broad-leaved and narrow, were expanding their foliage above the surface of the ponds and streams; and many flowers were in evidence all along the way.

25

Among our native shrubs in bloom, the spice-bush (*Benzoin*) and the shad-bush (*Amelanchier*) were conspicuous, while the cultivated trees were conspicuously represented by the apple (*Malus*), the pear (*Pyrus*) and the peach (*Amygdalus*). The herbaceous plants that attracted the eye readily along the way in their respective habitats were bloodroot (*Sanguinaria canadensis*), buttercups (*Ranunculus*), fire-pink (*Phlox subulata*), may-apple (*Podophyllum peltatum*), false meadow-rue (*Syndesmon thalictroides*) and various naturalized weeds that grew in cultivated fields and along the fence rows.

South of Philadelphia, and particularly south of Washington, vegetation appeared much more advanced. There, for several hundred miles, the Japanese honeysuckle (*Nintooa japonica*) was the naturalized plant most in evidence. This often much-abused vine seems to be excellent for holding sloping road-sides and railroad embankments; and, in addition to being evergreen or nearly so, it usually has a quite continuous flowering season, or several flowering seasons, except in mid-winter.

In western Delaware, pine trees had appeared; first the poverty-pine (*Pinus virginiana*), then the pitch-pine (*Pinus rigida*). The most conspicuous evergreen shrub accompanying these in the fall-line hills of Maryland was the calico-bush or mountain-laurel (*Kalmia latifolia*). Although not in flower, the numerous deep-green leaves, of the same fresh glossiness all the year round, made the plant notable. Thence, on down into Florida, one may observe not less than seven additional kinds of pines, or nine kinds altogether. It was always striking, where pine trees and broad-leaved trees grew associated, to see how different were the greens they presented—the greens of the broad leaves so bright and, as one might say, joyous; those of the pines so dull and sombre.

In Virginia and the Carolinas spring had progressed apace. Here meadows and pastures often supported so copious a growth of dandelions (*Leontodon Taraxacum*) that the myriads of flower heads suggested a carpet of gold; while other pastures were so thickly overgrown with a winter-cress (*Barbarea*) that, be-cause of the particular yellow of its flowers, they seemed sheathed in brass. Also there were old fields here and there that seemed to be covered with a bluish mist, an optical illusion, caused by the myriads of the small blue flower of the toad-flax (*Linaria canadensis*).

Thus we took a fleeting wild-flower census: violets (*Viola*), dewberries (*Rubus*), phlox (*Phlox*), wild-strawberries (*Fragaria*), pitcher-plants (*Sarracenia*), fleabanes (*Erigeron*), bladderworts (*Utricularia*), Virginia-cowslip (*Mertensia*), sneeze-weed (*Senecio*), flags (*Iris*), beard-tongues (*Pentstemon*), false-indigo (*Baptisia*). These were the more conspicuous. But also the woods were gay with dogwood (*Cynoxylon*), sweet-bay (*Magnolia*) and viburnums. Here, too, the ground was covered, often by the acre, with a beautiful white or pinkish azalea usually not more than a foot tall.

That old favorite of the South, the cultivated China berry (*Melia Azaderach*), was in bloom; and curiously enough, several weeks later I found it in the same condition in tropical Key West.

Nor were the palms wanting in this plant panorama, once we were in South Carolina. First we saw palmetto (*Serenoa serrulata*), then the blue-stem (*Sabal Adansonii*), and finally the cabbage-tree (*Sabal Palmetto*). With the exception of the needle-palm (*Rhaphidophyllum Hystrix*), the above-mentioned palms represent all the kinds that are found north of peninsular Florida. In Florida, however, nine other kinds grow naturally, these often represented by countless individuals.

Two interesting conditions were observed in passing through the various swamps. First: in the numerous swamps away from rivers, the early flowers of the spatterdock or bonnets (*Nymphaea*) and the water-lily (*Castalia*), seemed to be smaller than the later flowers, a condition usually, if not always, the reverse with our common herbs, where the early flowers are found conspicuously larger than the succeeding ones. Second: in the river swamps or bottoms, or the dry or partly dry plain between the river and the former banks of the river, at ordinary low water, hickories (*Hicoria*), oaks (*Quercus*), gums (*Nyssa*), maples (*Acer*), and elms (*Ulmus*), as well as the cypress (*Taxodium*), all develop conical bases or buttresses, evidently so as to make a more extensive root system for anchorage in the alluvium, particularly in seasons of high water.

Also, in depressions in the districts of little elevation, we observed groups of trees of the black-gum (*Nyssa biflora*) similar to those of the pond-cypress (*Taxodium ascendens*) in similar situations called cypress-ponds. The ponds with these distinctive groups of gum-trees may be called "gum-ponds." An investigation of their flora would doubtless be interesting.

The most attractive spring-flowering plant in all the swamps is the native wisteria (*Krauhnia frutescens*). This is a woody vine that climbs into the shrubs and trees and bears numerous drooping clusters of beautiful blue flowers. This species and another, which occurs in the Mississippi Valley, are the only representatives of the genus in the New World. The other species occur in eastern Asia.

With northern Florida, the chromatic scale and seasonal sequence of "the flowers that bloom in the spring" came to an end. The Florida flat-wood and the minor plant-regions have rather little to show, at a distance at least, with the changes of the season and latitude.

We reached Miami during the last week of April. The greater part of May was spent in the field. I was accompanied by Mrs. Small, who cared for the herbarium specimens incidentally collected. With the permission of Dr. David Fairchild, we established our field-headquarters in the laboratory building of the Plant Introduction Garden of the United States Department of Agriculture, in Miami.

Experience in the field and information gathered from time to time indicated five promising regions in Florida for discoveries in the genus *Harrisia:* (1) the sand-dunes along Saint Lucie Sound; (2) the islands of Tampa Bay; (3) the Cape Sable region, including the Madeira Bay district to the east and the Ten Thousand Islands to the north; (4) the Upper Florida Keys; (5) the Lower Florida Keys.

We took the field almost immediately after reaching Miami, and thereafter our active field-day, both on land and on water, almost invariably lasted from sunrise to midnight.

Miami and Terra Ceia Island

We first set out for the mouth of the Manatee River, our objective being Terra Ceia Island in McGills Bay. Our course lay along the eastern coast of the State as far north as either Jupiter or Fort Pierce. Various plant associations were traversed over this course.

First, the pine-woods are dominant. Of course, creeks with their bordering hammocks or prairies were crossed from time to time.

Then, north of the settlement of Hallandale, the "scrub" appears. This plant association occupies a formation of almost snow-white sands which are evidently ancient, now stationary, dunes. The southernmost dunes are low and nearly flat, but northward they become more undulating, and finally a few miles north of West Palm Beach, become conspicuous hills. Yet, to one's great surprise, among these hills, which attain a considerable elevation for that region, are situated a number of large cypress heads, some of which extend nearly or quite to the coast.

The characteristic tree of the "scrub" is the spruce-pine (*Pinus clausa*), but in its southern extension one finds a mixture of the Caribbean-pine or slash-pine (*Pinus caribaea*) and the sand-pine, or sometimes the Caribbean-pine exclusively. The conspicuous shrubs of the "scrub" are the saw-palmetto (*Serenoa serrulata*), the rosemary (*Ceratiola ericoides*), the prickly-pear (*Opuntia ammophila*), a heath (*Xolisma speciosa*), and scrub-oaks (*Quercus*). Various, but not numerous, herbs, both coarse and delicate, are associated with the shrubs, and vines are present. Among the latter a parasite (*Cassytha*), resembling the dodder in habit and color but really related to the laurels, is by far the most conspicuous. It commonly grows so luxuriantly that it actually smothers the shrubs, sometimes over large areas.

In the marshes among these dunes, and in and about the ponds, there was an abundance of aquatic and semiaquatic plants chiefly with yellow flowers. A score of yellow-flowered plants might be mentioned to one showing a different hue. One of the more conspicuous, although in no way showy plants was a tall sedge (*Dichromena latifolia*). This grew scattered or in colonies. The stalk was inconspicuous or invisible from a distance; but the whorls of long ghostly-white bracts at the top of the stalk were very conspicuous and suggested, especially toward dusk, so many little will-o'-the-wisps. It is one of the few sedges in which fertilization is accomplished through the agency of insects.

The old settlement of Jupiter, eighty-three miles north of Miami, was soon reached. There we had to decide whether we would strike out for the Okeechobee region over the old Fort Bassinger trail which was used during the Seminole Wars, or drive further north to Fort Pierce to enter the Okeechobee country. Inquiries not bringing forth any definite information, as usual, we decided, as we had already been over the Fort Pierce route, to try the Fort Bassinger trail, with Okeechobee City fifty odd miles to the northwest and Fort Bassinger about twenty miles further on along the Kissimmee River.

For the distance of a few miles the trail had been improved by surfacing the more or less improved grade. We passed beyond the "scrub" and the high pinelands, then crossed some streams with rich bordering hammocks.

We now approached "Hungry Land," so-called, they say, because a herd of stolen cattle were here penned up and allowed to starve, when the frightened thieves fled. But, in any case, it is well named. It is a desolate region east of the Saint Lucie slough; not like the general surface of the earth, one-third land and two-thirds water, but—well, to be charitable—say half land, half water. The more conspicuous plant associations were cypress-swamps, ponds, prairies, and low pine woods.

The ponds were partly dried up, but only to be so for a few days, as will be noticed on a subsequent page. Most of them, both as to the dry edges and the shallow water, supported a marvelously copious growth of the erect bladderwort (*Stomoisia juncea*). Acre after acre was almost a pure growth of the plant, inconspicuous in itself, but very conspicuous in such masses.

Another phenomenon that attracted our eye particularly was the numerous small pine trees, two or three years old, springing up among the cypress trees in the cypress swamps. They seemed out of place. Perhaps the lowering of the general water-table, consequent upon the decided lowering of the water of Lake Okeechobee, has brought about conditions permitting these pine trees to grow where formerly it was too wet.

After several hours in Hungry Land we came to the Hungry Land slough, which seems to represent the boundary line between two geological formations, namely: the Palm Beach Limestone and the Pleistocene. This slough, with its extensive hammocks, geographically separates Hungry Land, which lies between it and the Atlantic Ocean, and the Allapattah Flats, which lie between it and Lake Okeechobee.

We crossed the slough, which has lately been dredged as part of the Saint Lucie Canal, on a ferry, and entered the Allapattah Flats. Here the ground soon became noticeably higher than that of Hungry Land, and the landscape was enlivened with bush clovers (*Petalostemon*), devil's shoe-strings (*Galactea*), thyme (*Pycnothymus*), grass-pinks (*Sabbatia*), asclepiads (*Asclepiadora*), bonesets (*Eupatorium*), goldenrods (*Solidago*), asters (*Aster*), and thistles (*Cirsium*).

These "flats" have long been a rendezvous for the Cow Creek Indians and we soon could discern many camps in the distant pine wood towards the west and an occasional Indian walking, or riding on horseback through the woods. The land continued to rise and the pine trees became more scattered. Then open stretches appeared, and, finally, the trail led out to an extensive prairie. Far to the west we saw a long, evidently tall hammock, but in this flat country one would have almost sworn that this was a range of hills.

We had not time to stop for an examination of the flora of the prairie, but two shrubs were particularly noticeable. Parts of the prairie were covered with a dense growth of the waxmyrtle (*Cerothamnus*), the bushes ranging from one to three feet tall. Other areas were covered with a small oak, usually only six to ten inches tall—this lack of visible development probably made up for by an extensive growth underground.

At a fork in the trail we decided to take the left hand or westerly branch, as it seemed to lead toward the distant hammock. Although we had several times seen this hammock from the other side and had been in it, its identity at this time did not dawn upon me. On approaching it we saw that the trail led through it. When part way through the timber we could look beyond the cut into space, and a little further on a vast sheet of water appeared. It was Lake Okeechobee! We drove out onto a beach which only a few years ago was still the bottom of the lake. More than that, we were soon actually driving over the very course we had sailed over five years previously[1] in a forty-five foot cruiser!

But how much everything had changed since that time! Instead of a natural beach close to a primeval hammock, we

[1] Journal of the New York Botanical Garden 15: 69-79. 1914.

found several hundred yards of exposed new weed-clothed lake-bottom, down from the old beach line; and as for the hammock, it was wrecked. Fire had been in it, perhaps more than once; and, in many places, instead of the once magnificent verdure, one saw only dead giant cypress trees, standing desolate, or prone in the wholly or partly burned humus where once had thrived an almost impenetrable mass of ferns and other herbaceous plants.

Sixteen miles over the former lake-bottom and through a portion of the destroyed hammock brought us to Okeechobee City.

Early the following morning we set out for Bradentown, nearly two hundred miles further on. We continued on the Fort Bassinger trail to the Kissimmee River, and thus on to Fort Bassinger itself, a course we had already traversed in the opposite direction one night last December.[1] Covering it in daylight, however, gave us quite a different impression of the region. The scattered pine trees north of Lake Okeechobee soon gave way to open prairie flanked in the far distance by pine-woods or hammock, east and west. The higher parts of the prairies were mostly grass-clothed, while the lower portions were clothed with almost a pure growth of flag (*Iris*) as far as the eye could see. The flags were mostly past flowering and bore clusters of cucumber-like capsules, the weight of which had borne the weak flower-stalks to the ground.

Seldom did the trail fail to yield something of interest. About half-way to Fort Bassinger we passed a cypress-head which served as rookery for a flock of wood-ibis, and many of these beautiful birds were roosting on the trees, shining whitely against the shadows.

It was midway to the Kissimmee River that growth of scattered pine trees, small cabbage-tree hammocks, and live-oaks appeared, albeit the land continued low. Further north there was an extensive growth of myrtle (*Cerothamnus*), gallberry (*Ilex*), heath (*Xolisma*), and scrub-oak (*Quercus*), as well as considerable turf. On approaching the Kissimmee we found curious circular areas of saw-palmetto (*Serenoa serrulata*) and

[1] Journal of the New York Botanical Garden **21**: 25–38, 45–54. 1920.

small persimmon trees (*Diospyros Mosieri*)[1] ,and groves of live-oaks that looked almost like apple-orchards.

Then after crossing the wide flood plain of the Kissimmee we came to the shallows and blind channels of the river. These were choked or carpeted with dense growths of aquatic grasses (*Panicum*), pond-weeds (*Potamogeton*), naiads (*Naias*), and arrow-heads (*Sagittaria*).

The settlement of Bassinger lies east of the Kissimmee, while Fort Bassinger proper, of Seminole War times, was on the western side of the river.

Crossing the river west of Bassinger, we left the Okeechobee prairie and were soon on the Istokpoga prairie, traveling east and north of Lake Istokpoga, a little-known body of water about ten miles in diameter. We then crossed Istokpoga Creek, which connects the lake of the same name with the Kissimmee River, and is apparently the only outlet of the lake. It is lined with beautiful hammocks. The dense hammock about Lake Istokpoga itself, plainly visible in the distance, is a botanical storehouse, as yet wholly unexplored. We do not even know whether its arboreous growth is coniferous (cypress) or broad-leaved, or both. After circling many miles toward the west and crossing a creek flowing from Lake Arbuckle into Lake Istokpoga, we met the pine woods again.

A meandering trail soon led us to a most insignificant stream apparently, but this stream proved to be the boundary between two conspicuously distinct geological formations; for thence the trail entered immediately a series of curious old sand-dunes wholly unlike the lands we had left. These dunes ran in somewhat parallel ridges, now close together, again separated by extensive "parks"; and the sand, although snow-white and loose, supported a veritable flower garden. In general, the region forms a watershed, the streams on the eastern

---

[1] **Diospyros Mosieri** n. sp. A tree seldom as large as *D. virginiana* or commonly shrub-like; differs from *D. virginiana* in the smaller flowers, the staminate only about one-half as large, the short and broad calyx-lobes, the stouter stamens, and the short, broad, turgid seeds. The type specimens were collected in pine woods near the Humbugus Prairie, west of Little River, Florida, in fruit, by J. K. Small, C. A. Mosier, and G. K. Small, No. 6927, July 8, 1915; in dead ripe fruit, by C. A. Mosier, November 1917, and in flower, by J. K. Small, April 1920.

side flowing into the Okeechobee basin, those on the western side forming tributaries of the Pease River. Curiously enough, the backbone of this plateau has a chain of large and small lakes on it, the magnet of several settlements.

After crossing a series of other ridges we came to Sebring. Thence we traveled northward on the plateau to Avon Park. Another leg of our course westward brought us into the valley of the Pease River at Zolfo Springs.

In the intervening country we found small forests of the persimmon in full flower. South of Zolfo Springs terrestrial orchids (*Limodorum Simpsonii* and *Gymnadenopsis nivea*) appeared in the marshes for the first time since we left the eastern Okeechobee region. As for the pine woods, the conspicuous plant here was a curious yellow-flowered legume called *Chapmania*. This has a deep-seated, woody root and a slender wand-like stem, bearing yellow flowers whose petals do not at first sight suggest a papilionaceous corolla at all. It was named for Dr. A. W. Chapman,[1] long a resident of Florida, and the first botanist to construct an interpretation of the flora of the southeastern United States. The hammocks of the valley were fragrant with the sweet-bay (*Magnolia virginica*) and numerous herbs and shrubs were in bloom, notably the tall blackberries (*Rubus*) and the wild purple leather-flower (*Viorna crispa*). Following the valley southward for about twenty miles, we arrived at Arcadia.

Although we had traveled many miles to the west, we were still over forty miles in a bee-line from the coast of the Gulf of Mexico. Incidentally, the intervening country is essentially unsettled and is accessible by a very poor trail.

We soon crossed the Pease River and once again struck into the wilderness. Two settlements are on the trail between the Pease River and Sarasota on the Gulf, according to the map,

[1] Alvan Wentworth Chapman (1809-1899) was a graduate of Amherst College who went to Georgia as a teacher, studied medicine there, and soon afterward entered upon the practice of his profession in Florida. For more than half a century his home was at Apalachicola, where he died in his ninetieth year. His "Flora of the Southern United States," first published in 1860, and running through several editions, was for nearly fifty years the only manual of the flowering plants of the southeastern states.—JOHN HENDLEY BARNHART.

but these are barely visible to the naked eye. The region is wild and lonely, bisected by the Myakka River with its wide prairies. On either side of the prairies are unpeopled pine woods. And through all of this runs a trail that is mostly sand, sand that is very loose and often deep. The sand was various in color; and there, instead of the snow-white sand form i g ridges, it filled the low places. But all of it was hard to travel through. At many places, to use the language of the country, our motor seemed "powerful weak." Still, we did manage to crawl along. Night was coming on.

Several miles beyond the tiny settlement of Myakka, when it was dark, we found that we had missed the trail and that we were on some sort of an indigenous branch trail, so common to the country. Perhaps the branch would lead back into the main trail, as branches often do. We continued. It was very dark, but the stars were out. By the stars we could tell that we were traveling not west, toward our destination, but south into a still more savage wilderness. We persevered, and at last were rewarded, or so we thought, by seeing several bright lights to the westward, then yet another branch trail that would lead us there.

But our lights—alas!—merely betokened a forest fire.

The fire was extensive. It had curiously run into about a dozen lines, and each of these lines cut the trail with a hurdle of flame. But we could not, or would not, turn back. We took some photographs, then beat out enough of the flame to make a safe passage for our extra supply of gasoline, and went ahead.

Several miles of meandering through the dark woods brought us eventually back into the main trail, and thus, in due time, other and less sinister lights appeared ahead. They were the lights of Sarasota. After that, plain sailing, the highway to Manatee, another's hour's run, and we were in Bradentown, our objective. An eleven o'clock supper, then bed.

According to prearranged plans, the following morning found a motor boat ready to carry us to McGills Bay and Terra Ceia Island. Before we started we were joined by Mr. Alfred Cuthbert, who generously entertained us during our stay in that part of the country.

Cactus plants, an abundance of prickly-pears (*Opuntia Dillenii* and *O. austrina*), came into evidence as soon as we landed on Terra Ceia Island, but these were not what we sought. While wandering about in the woods we came upon a cabin, and a girl who lived there said she thought we would find other cacti further up the shore. Also, she volunteered to guide us past the "dead bodies" in the trail! We followed. The bodies were there, sure enough, but only skeletons, happily. They were the fossil remains of aborigines dug from a shell midden in making a drainage ditch.

Terra Ceia Island itself is a vast kitchen-midden, or ancient artificial shell heap, built up by the former inhabitants with their discarded oyster, clam, and conch shells. The bones we saw, evidently strongly impregnated with lime, were in a good state of preservation. The skulls were particularly interesting. Many of us have been taught that a diet of sea-food, particularly develops the brain. Now these ancient people lived largely, perhaps wholly, on fish and "shell-fish"; but from what we know of their history they were not intellectual giants. Their diet did evidently develop their heads, however. The skull-bones of the specimens we observed varied from one-half an inch to nearly three-quarters of an inch in thickness.

After this venture in archaeology, we broke into the hammock at several points and soon found our prize, a tall columnar night-blooming cactus with stout stems growing to a height of twenty feet. It proved to be a new species of *Harrisia*,[1] differing from all the other known Florida kinds in having brown wool and yellow fruit, the other kinds having white wool and red fruit. Fortunately, we secured one flower-bud about to open and one ripe fruit. The fruit was mailed directly to the Garden for painting, while the bud was taken to Mr. Cuthbert's house, where we sat at the festive board from six o'clock till midnight, shortly before which the bud opened and we secured a description of the flower. More closely than in any of the other Florida species, the flower-limb resembles the expanded flower of a water-lily.

---

[1] This species was named for the prehistoric inhabitants of that region, *Harrisia aboriginum*, and published in THE CACTACEAE **2**: 154. 1920.

PLATE 254

On Punta Blanca at mouth of the Caloosahatchie. Fire-place of the aborigines in foreground; dense hammock in background. The original surface of the kitchen-midden) as the top of the rock, the remains of the fire-place. The loose material, mainly shells a aborigines, was removed for road-making. Continuous fires of generations burned the shells and the action of the elements concreted the residue into a solid mass of lime. Since their disuse by the natives the shell heaps have become covered by a dense hammock-jungle.

A botanical survey of Terra Ceia Island would doubtless prove interesting in other respects. Its flora has more tropical elements than any of the neighboring islands. It is probably the northern limit of the Gumbo-limbo (*Elaphrium Simaruba*) on the western coast, which species is represented there by a group of very large trees. Armed shrubs were much in evidence. The cat's-claw (*Pithecolobium*), the wild-lime (*Zanthoxylum*), the nicker (*Guilandina*), the devil's claws (*Pisonia*), the prickly-pears already mentioned, and other spiny plants often made the hammocks almost impenetrable.

Having accomplished our purpose we set out for headquarters, Miami, early the following morning. Instead of returning by the route we had just been over, however, we set out in the opposite direction, heading for the Tampa region. One reason for our selection of this route was a possible visit to the fern-grotto described in a former paper,[1] and which should have an interesting spring flora, besides being a remarkable fernery. But after driving north of Plant City for several miles we found the highway in such abominable condition that lack of time forced us to abandon this itinerary. Somewhat discouraged, we turned back and started through the southern end of the lake region.

West of Plant City we observed several ridges of "scrub." East of it were hammocks with oaks (*Quercus*), hickories (*Hicoria*), and red-gum (*Liquidambar*), prominently developed. Of course, there were pine woods and a mixture of pine woods and oak woods.

Evening found us again at Avon Park, which is situated in a very interesting floral region.

One locality just west of the town attracted our attention particularly. It was a park-like area between hills. The ground was thickly covered with an intermingled mass of a morning-glory (*Stylisma*) and toad-flax (*Linaria floridana*). The conspicuous shrubs at the height of their flowering season were a large-flowered prickly-pear (*Opuntia ammophila*) and the largest-flowered papaw (*Asimina obovata*). The adjacent hills were, moreover, in places yellow with the numerous flowers of a rare false-indigo (*Baptisia LeContei*).

[1] Journal of the New York Botanical Garden **21**: 25–38, 45–54. 1920.

The following morning we continued our journey eastward and soon came to the crest of the sand-plateau referred to in a preceding page. There we made our first extended stop for collecting miscellaneous plants. The pine of those ancient dunes was not the spruce-pine (*Pinus clausa*), the normal tree of such dunes, but the long-leaf pine (*Pinus palustris*). The broad-leaved trees were mostly represented by the turkey-oak (*Quercus Catesbaei*) and a kind of hickory (*Hicoria*). Shrubs were more numerous than trees, both in kinds and quantity. In this category may be mentioned the scrub-plum (*Prunus geniculata*), our smallest and latest discovered eastern plum, at this time not only past the blooming season, but also without fruit. Associated with the plum was a curious kind of titi (*Cyrilla*) just coming into flower. The only two genera of the Yucca family east of the Mississippi River were represented by *Yucca filamentosa* and *Nolina atopocarpa*. Dozens of different kinds of herbs were in full flower with a great range of colors, but the most showy plant, both in quantity and color, was the blue lupine (*Lupinus diffusus*). This plant was in evidence everywhere in high pineland and was particularly conspicuous on account of its silvery foliage and sky-blue flowers. The most interesting vine of the sand-hills was the small bullace-grape or muscadine (*Muscadenia Munsoniana*). This widely distributed grape of southern Florida usually has leaves from two up to four inches in diameter, but the plants we found on the sand-hills had leaf-blades almost uniformly with a diameter of less than an inch.

New plants were not wanting. There were not only new species, but a new genus was represented by an abundance of a prostrate shrub.

One particularly attractive new species of uncertain relationship, with the habit of a yellow-flax (*Cathartolinum*), was abundant in the white sand. The new genus to which reference has already been made[1] was found with both flowers and ripe fruits. It belongs to the knotweed family and is related to the genera *Polygonella* and *Thysanella*.

Examinations of those ancient dunes at successive seasons would doubtless yield additional novelties, for the seasonal

---

[1] Journal of the New York Botanical Garden **21**: 48. 1920.

periods of the plants, particularly annuals, under the desert conditions prevailing there, are evidently exceedingly short.

A spiderwort (*Tradescantia*), perhaps new, should be mentioned here on account of the delicious violet-fragrance of its bright-blue flowers. It was abundant in the high pineland.

Leaving the sand ridges and continuing eastward we were soon back on the Istokpoga prairie. And here, in one low place, we found a veritable orchid garden. Two terrestrial orchids, a snake mouth (*Pogonia*) and a grass-pink (*Limodorum*) grew amidst the low grass in countless thousands.

Fire! Fire!! Fire!!!

This word might be repeated scores of times during a day's travel in southern Florida. In nearly every direction one turns clouds of smoke go rolling skyward. On this occasion nearly the whole Istokpoga prairie was on fire. And far to the southeast was that continuous cloud of smoke from the delta of the Kissimmee River. A doleful story, for the hammock and humus there has been burning for years, the heaviest rains having failed to extinguish it.

Observation on the Kissimmee flats shows the herbaceous vegetation to be composed of very low, often prostrate, plants. Those blooming at the time we were there were two kinds of figworts, a hedge-hyssop (*Gratiola*) and a false-pimpernel (*Ilysanthes*). The shrub, scattered over the flats, was a bright-yellow St. John's-wort (*Hypericum nudiflorum*). We had noticed the trunks of the cabbage-trees (*Sabal Palmetto*) in the vicinity of the flats to be broadly girdled about three feet above the ground, and were at a loss to account for the condition. However, a little observation soon solved the problem. The cattle of the region, being unable to rid themselves of flies collected on their sides out of reach of tail or tongue, walk around the cabbage trees scraping their sides against the rough trunk, thus getting rid of the flies, but gradually wearing the tree away.

As the afternoon was well spent, we hurried across the prairie to the head of Lake Okeechobee, thence down the eastern shore as the sun set, and out on the prairie east of the lake just before darkness fell. Thence we made our way over the trail to the eastern coast as fast as possible. In the fifty odd miles to Jupiter we saw no signs of a human being except the ferryman at the Saint Lucie slough, the lights of a few lone houses nearby, and the campfires of scattered Seminole Indians here and there in the pine woods. We reached West Palm Beach at midnight and spent the remainder of the night there. Early the following forenoon found us in Miami.

(To be continued)

THE PRESERVATION OF OUR NATIVE PLANTS

The income from the Stokes' Fund for the investigation and preservation of our native plants has not been used since 1916 for any further publication of colored plates, prices of printing having greatly increased and good results in color-work being difficult to obtain. There are four more plates with descriptions ready to print of the Canada Lily, Cardinal Flower, Holly, and Rhododendron.

But the time has not been lost, for we have added greatly to our collection of colored lantern slides, notably by the purchase of a beautiful set from the J. Horace McFarland Company; others have been received from Miss Elsie Kittredge, Miss Fleda Griffith, and from L. W. Brownell. They have been used to good advantage for the benefit of the Garden Clubs, two of whose objects are: "to aid in the protection of native plants and birds; and to encourage civic planting."

Lectures have been given to the following local Garden Clubs: Staten Island, New Rochelle and Larchmont, Bedford and Pleasantville, New York; Leonia, Short Hills, Plainfield, Bernardsville, Rumson, Princeton and Trenton, New Jersey; Litchfield and Ridgefield, Connecticut; and Lenox, Massachusetts. To all of these, duplicates of our leaflets, pictures, posters, buttons and pledges have been sent and a number of them have joined the Wild Flower Preservation Society as Associate or Life Members. "Wild Flower Committees" are being organized this year by many of the Clubs that are members of the Garden Clubs of America and they are planning to devote "one program a year to wild flowers and native conservation" and are giving space in each issue of their Bulletin to this department. Two leaflets have so far been issued: number 1, on "State and County Parks," being an account of an interview with Henry C. Cowles, Professor of Plant Ecology of Chicago University, President of the Wild Flower Preservation Society and a member of the Committee on "Plant Sanctuaries" for the Preservation of Natural Conditions, of the Ecological Society of America, who in coöperation with Professor Bray, of Syracuse, J. W. Harshberger, of the University of Pennsylvania, and others, have been devoting much attention to this work since 1917. The Committee includes sixteen other members and their records cover all the States of the Union and British America with advisers in Entomology, Ornithology, Fisheries and Grazing.

Having served also as Secretary-Treasurer of the Wild Flower Preservation Society, for several years, the correspondence has been large and more or less similar, with numerous requests for information, literature and help. We feel greatly encouraged by the fact that an active Chapter has been founded in Chicago with seven hundred members and twenty-six associate schools; they are coöperating with about thirty-five of the local clubs of Chicago and its vicinity, including Botanical Clubs and Audubon Societies, by holding annual exhibits covering many branches of Natural History and educating the children to be interested in all of them! They also give Summer Fêtes for the children.

In Washington, where some of our earliest work was done, they have reorganized and are planning for Arbor Day work this year. They have issued circulars to teachers for use on Arbor Day and May Day, have given lectures and exhibits and conducted field trips for the Boy Scouts for observing and photographing wild flowers. A printed series of posters of the Maryland State Laws, protecting the Holly, Mistletoe and Evergreens are issued free and they have also conducted a publicity campaign in the newspapers of the District of Columbia.

We have Associate Clubs and members in each of the following states: New York, New Jersey, Pennsylvania, Maryland, District of Columbia, Florida, Texas, Arkansas, Illinois, Indiana, Ohio, Wisconsin, Minnesota, Michigan, Colorado, California, Washington and Vancouver Island. We are also a member of the Mountaineering Club of America and receive many of the publications of the Associated Clubs.

Our President, Dr. Cowles, has been a delegate to the National Parks Convention at Madison, Wisconsin, January 11 and 12, where he represented the Wild Flower Preservation Society and the Ecologists of America.

The Ecological Society of America, organized in 1914, has been studying the effects of climate and weather and of soil on the association of plants into natural groups or plant formations; their geographic distribution and relation to animals, including insects, birds and mammals; the sequence of the reappearance of vegetation in devastated regions, and the preservation of natural conditions for both animals and plants. They include among their members: Zoologists, Botanists, Entomologists, Foresters, Geographers, Plant Pathologists and Naturalists.

We have also had some correspondence and exchange of publications with the Swiss League for the Protection of Nature and received a set of the auto-chrome plates of their Alpine Gardens, full of beautiful wild flowers. Through the Smithsonian Institution we have also supplied a full set of our publications to the Czechoslovak Republic. Even the Botanical Survey of the Union of South African States is fostering an interest in their native plants in order to encourage their preservation and cultivation!

The National Geographic Society has published forty-four of Miss Eaton's exquisite colored plates of native and introduced plants and an account, also illustrated in colors, of the State Flowers. In May, 1915, they published an article on the May Flower, by F. V. Coville, and June, 1916, another by Mr. Coville on the Taming of the Wild Blueberry, both of great interest.

The Comstock Publishing Company of Ithaca, New York, has recently issued a set of twenty-nine uncolored plates, also from Miss Eaton's drawings, for use in schools, including twelve of our wild flowers: Trailing Arbutus, Hepatica, Dutchman's Breeches, Bloodroot, Spring Beauty, Adder's Tongue Lily, Columbine, Marsh Marigold, Violet, Trillium, Moccasin Flower and Jack-in-the-Pulpit. It is a most encouraging sign that everywhere efforts are being made to reach the school children, boy scouts, camp-fire girls and girls' camps, and Young Men's and Young Women's Christian Associations, and to coöperate with the Audubon Society in trying to teach the children to love the birds and wild flowers and to help preserve them, and by means of poems, drawings and illustrated lessons they are becoming familiar with many species. The Nature Study League of New York City, under the able guidance of Mrs. Northrop and the Natural Science Committee of Hunter College, is reaching some of the down-town schools of Manhattan, and is holding its first public meeting at the American Museum of Natural History on February 9, 1921.

We have lectured and sent pledges, buttons and the Stokes' sets of colored pictures of wild flowers needing protection to many of the public schools, high schools and colleges and public libraries of the Bronx, Manhattan and Brooklyn. We have also coöperated with the Conservation Department of the General Federation of Women's Clubs of which Mrs. John Dickinson Sherman is Chairman and sent them an exhibit of our publications and pictures, and have also sent all our publications to the New York State Conservation Commission.

In Bulletin 9 of the "Permanent Wild Life Protection Fund," Mr. Hornaday has a report on the progress of the movement to induce land-owners to forbid shooting of any kind on their estates. There are now no less than 3131 such sanctuaries in the United States, covering in all a million and a half acres. Oregon leads with over 800,000 acres and yet Oregon in a referendum vote has just declined to preserve the famous Malheur Lake bird reservation and has turned it over for land drainage and speculation. It is hoped that through the influence of the Garden Clubs of America and the Ecological Society, similar reservations for the protection of plants may be secured and the cooperation of all nature lovers is solicited.

In May, 1917, the American Museum Journal devoted three pages to an article on The Conservation of Wild Flowers, which Miss Dickerson illustrated by a series of twelve photographic reproductions from native plants and they also issued a beautiful series of duotone photographs with an article on the Season of Wild Flowers. The illustrations were done by the De Vinne Press of New York and are exquisite examples of fine reproduction.

The importance of many native plants as food for birds and their relation to insects and agriculture has not been overlooked, and Dr. G. Clyde Fisher, of the American Museum of Natural History, delivered a lecture here at the Garden on May 18, 1920, and is preparing a leaflet for publication on this topic. He has also given a well-illustrated lecture on the relation of forests to rainfall and water supply.

Mr. Norman Taylor, of the Brooklyn Botanic Garden, whose practical knowledge of the local flora is of great value, has also been coöperating with us by delivering some of the lectures to the Garden Clubs of the vicinity, and it is hoped that he and Mr. Durant, who is associated with Mr. Gillette, will tell us how to transplant and grow the wild flowers successfully.

The Torrey Botanical Club, whose membership is mostly in New York State, has become a Life Member of the Society, and two of our annual meetings have been held in conjunction with theirs, our State Botanist, Dr. Homer D. House, having addressed us last year and exhibited the colored plates from the Wild Flowers of the State, recently published. Dr. House has been exerting his influence at Albany and throughout the State to arouse public feeling, and Professor W. L. Bray has also been interested in the preservation of the Hart's Tongue Fern at Syracuse, and in coöperation with the Syracuse Botanical Club and the Ecological Society of America, has succeeded in having set aside the Station at Green Pond as a State Park.

ELIZABETH G. BRITTON,
New York Botanical Garden,
Bronx Park, New York City

FRANCIS LYNDE STETSON

Francis Lynde Stetson, a member of the Board of Managers of the New York Botanical Garden since 1908 and a Vice-President of the Garden since 1914, died at his residence in New York City on December 5, 1920, after a prolonged illness. He was in the seventy-fifth year of his age.

Throughout his association with the institution he was active in its behalf, serving nearly continuously as a member of the Executive Committee and for a series of years as its legal adviser. He was keenly interested in all phases of its work, more particularly, perhaps, in its scientific and educational development, but he was intensely fond of plants of all kinds, a born nature-lover. His large estate at Skylands, in northern New Jersey, his summer home, gave him great pleasure, providing space for the planting and cultivation of a noteworthy collection of trees and shrubs, with vegetable and flower gardens and extensive farming operations, together with ready access to large areas of forests, fields and meadows, and he took keen delight in all these. Wild plants were of great interest to him and he was an enthusiastic advocate of the preservation of natural features and the conservation of natural resources.

He freely and generously contributed both time and money to the development of the New York Botanical Garden, and his advice has been of great value to the institution. His services to botanical science are commemorated by the genus *Stetsonia*, a gigantic and characteristic cactus of the Argentina Republic, named in his honor.

Resolved, That the Managers of the New York Botanical Garden deeply mourn the loss of an esteemed and beloved friend and associate.

Resolved, That the foregoing preamble and resolution be entered on the minutes, and that a copy be transmitted to his bereaved family.

Approved by the Board of Managers of the New York Botanical Garden January 10, 1921.

N. L. BRITTON,
Secretary

FANNY BRIDGHAM FUND

The Garden has recently received a legacy of $30,000 from the executors of the will of the late Mrs. Samuel Bridgham. At the annual meeting of the Board of Managers on January 10, 1921, the following resolutions were adopted:

Resolved, That the legacy of $30,000 received from the estate of Mrs. Fanny Bridgham be designated a permanent fund under the name "Fanny Bridgham Fund," and that its income be used, after investment, for the purchase and binding of books for the library unless otherwise ordered by the Board of Managers.

Resolved, That the managers gratefully receive the generous gift of Mrs. Bridgham and that the Director-in-Chief is authorized to express their appreciation of the gift to Messrs. Cadwalader, Wickersham & Taft, with the request that they communicate these resolutions to Mrs. John Innes Kane, sister of Mrs. Bridgham.

The following letter was received from Mrs. Kane:

January 20, 1921.

My dear Dr. Britton:

Messrs. Cadwalader, Wickersham and Taft have communicated to me the resolutions adopted by the Board of Managers of the New York Botanical Garden, concerning the legacy left to the "Garden" by my sister, Mrs. Bridgham.

Will you kindly express to the Board my appreciation of these kind resolutions and my gratification that it has seen fit to designate this legacy as a Permanent Fund.

Very sincerely yours,

(Signed)   ANNIE C. KANE.

The income of this fund will aid greatly in the building up of the library.

Mr. Samuel Bridgham, who died some years ago, was a botanical artist of distinction. He made many drawings for the late Professor W. G. Farlow, of Harvard University, and many for "Illustrated Flora of the Northern States and Canada."

N. L. BRITTON

GREENHOUSE LECTURES

MARCH AND APRIL, 1921

The regular public lectures in the Central Display Greenhouse, Conservatory Range 2, will be given this spring at 3:15 o'clock on *Sunday* afternoons, instead of on Saturdays as heretofore. Living plants will be used in illustration. Dr. Rusby will open the series with a talk on South American drug plants, and this will be followed by discussions of various other groups of plants, some of which are of immense importance economically.

March 20—"South American Drug Plants," Dr. H. H. Rusby.

March 27—"Bulbous Plants and How to Force Them for the Home," Dr. M. A. Howe.

April 3—"Fiber Plants," Dr. A. B. Stout.

April 10—"Milk-trees and Other Lactiferous Plants," Dr. W. A. Murrill.

April 17—"Air Plants," Dr. H. A. Gleason.

April 24—Desert Plants," Mr. G. V. Nash.

Conservatory Range 2 is situated at the eastern side of the Botanical Garden, north of the Allerton Avenue entrance. It is most conveniently reached from the Allerton Avenue Station on the White Plains Extension of the Subway from East 180th Street. Visitors coming by train to Botanical Garden Station should inquire at the Museum Building.

W. A. Murrill

CONFERENCE NOTES FOR JANUARY

The regular monthly conference of the scientific staff and students of the Garden was held on the afternoon of January 5, 1921. Brief reports were made by Dr. H. A. Gleason, Dr. J. H. Barnhart, Dr. Michael Levine and Professor R. A. Harper, regarding matters of botanical interest at or in connection with the annual session of the A. A. A. S. Following these, Mr. G. T. Hastings gave a very interesting report on "Succession of Algae in the Grassy Sprain Reservoir," an account of which, prepared by Mr. Hastings, will appear later in this JOURNAL.

A. B. Stout,
Secretary of the Conference

NOTES, NEWS AND COMMENT

Mr. Oakes Ames, Director of the Botanical Gardens of Harvard University, has recently published the sixth fascicle of his "Illustrations and Studies of the Family Orchidaceae," and has presented a copy of this important work to our library, the five preceding fascicles having been presented by him as they were published during the past few years. This fascicle contains two papers, the one on the Orchids of Mount Kinabalu, British North Borneo, in the writing of which Mr. Ames has been assisted by Mr. Charles Schweinfurth; the other is the seventh contribution by Mr. Ames upon Philippine Orchids. The fascicle is illustrated by twenty-one plates drawn by Mrs. Ames, and it is dedicated to Mr. Elmer Drew Merrill, in recognition of his distinguished services to Malayan Botany. The part of the fascicle on Philippine orchids is also published as a separate in twenty-five copies, one of which Mr. Ames has generously presented to the library of the Garden.

Over 600 biology pupils from Morris High School spent the forenoon of January 18 at the Garden studying the museum and living plant collections under the guidance of their teachers and members of the Garden staff. They also heard a lecture on forestry given by Mr. Inman.

About 200 biology pupils from Evander Child's High School visited the Garden on January 20 to study the plants in the Conservatories, the trees in their winter conditions, and certain economic plant products in the Museum. Mr. George T. Hastings delivered an address on forestry in the public lecture hall, which was much appreciated both by the pupils and their teachers. Members of the Garden staff acted as guides and instructors.

Meteorology for January.—The total precipitation for the month was 2.39 inches, of which 0.20 inches (two inches by snow measurement) fell as snow. The maximum temperatures recorded for each week were as follows: *59° on the 2nd*, 52° on the 4th, 53° on the 14th, 57° on the 23rd, and 44° on the 30th. The minimum, temperatures were: 20° on the 7th, 18° on the 13th, *4° on the 19th*, and 5° on the 25th.

PUBLICATIONS OF
The New York Botanical Garden

Journal of the New York Botanical Garden, monthly, illustrated, containing notes, news, and non-technical articles of general interest. Free to all members of the Garden. To others, 10 cents a copy; $1.00 a year. [Not offered in exchange.] Now in its twenty-second volume.

Mycologia, bimonthly, illustrated in color and otherwise; devoted to fungi, including lichens; containing technical articles and news and notes of general interest, and an index to current American mycological literature. $4.00 a year; single copies not for sale. [Not offered in exchange.] Now in its thirteenth volume.

Addisonia, quarterly, devoted exclusively to colored plates accompanied by popular descriptions of flowering plants; eight plates in each number, thirty-two in each volume. Subscription price, $10.00 a year. [Not offered in exchange.] Now in its sixth volume.

Bulletin of the New York Botanical Garden, containing the annual reports of the Director-in-Chief and other official documents, and technical articles embodying results of investigations carried out in the Garden. Free to all members of the Garden; to others, $3.00 per volume. Now in its tenth volume.

North American Flora. Descriptions of the wild plants of North America, including Greenland, the West Indies, and Central America. Planned to be completed in 34 volumes. Roy. 8vo. Each volume to consist of four or more parts. Subscription price, $1.50 per part; a limited number of separate parts will be sold for $2.00 each. [Not offered in exchange.]

Vol. 3, part 1, 1910. Nectriaceae—Fimetariaceae.

Vol. 7, part 1, 1906; part 2, 1907; part 3, 1912; parts 4 and 5, 1920; part 6, 1921. Ustilaginaceae—Aecidiaceae (pars). (Parts 1 and 2 no longer sold separately.)

Vol. 9 (now complete), parts 1–7, 1907–1916. Polyporaceae—Agaricaceae (pars). (Parts 1–3 no longer sold separately.)

Vol. 10, part 1, 1914; parts 2 and 3, 1917. Agaricaceae (pars).

Vol. 15, parts 1 and 2, 1913. Sphagnaceae—Leucobryaceae.

Vol. 16, part 1, 1909. Ophioglossaceae—Cyatheaceae (pars).

Vol. 17, part 1, 1909; part 2, 1912; part 3, 1915. Typhaceae—Poaceae (pars).

Vol. 21, part 1, 1916; part 2, 1917; part 3, 1918. Chenopodiaceae—Allioniaceae.

Vol. 22, parts 1 and 2, 1905; parts 3 and 4, 1908; part 5, 1913; part 6, 1918. Podostemonaceae—Rosaceae.

Vol. 24, part 1, 1919; part 2, 1920. Fabaceae (pars).

Vol. 25, part 1, 1907; part 2, 1910; part 3, 1911. Geraniaceae—Burseraceae.

Vol. 29, part 1, 1914. Clethraceae—Ericaceae.

Vol. 32, part 1, 1918. Rubiaceae (pars).

Vol. 34, part 1, 1914; part 2, 1915; part 3, 1916. Carduaceae—Anthemideae.

Memoirs of the New York Botanical Garden. Price to members of the Garden, $1.50 per volume. To others, $3.00. [Not offered in exchange.]

Vol. I. An Annotated Catalogue of the Flora of Montana and the Yellowstone Park, by Per Axel Rydberg. ix + 492 pp., with detailed map. 1900.

Vol. II. The Influence of Light and Darkness upon Growth and Development, by D. T. MacDougal. xvi + 320 pp., with 176 figures. 1903.

Vol. III. Studies of Cretaceous Coniferous Remains from Kreischerville, New York, by A. Hollick and E. C. Jeffrey. viii + 138 pp., with 29 plates. 1909.

Vol. IV. Effects of the Rays of Radium on Plants, by Charles Stuart Gager. viii + 278 pp., with 73 figures and 14 plates. 1908.

Vol. V. Flora of the Vicinity of New York: A Contribution to Plant Geography, by Norman Taylor. vi + 683 pp., with 9 plates. 1915.

Vol. VI. Papers presented at the Celebration of the Twentieth Anniversary of the New York Botanical Garden. viii + 592 pp., with 43 plates and many text figures. 1916.

Contributions from the New York Botanical Garden. A series of technical papers written by students or members of the staff, and reprinted from journals other than the above. Price, 25 cents each. $5.00 per volume. In the ninth volume.

NEW YORK BOTANICAL GARDEN
Bronx Park, New York City

Vol. XXII   March, 1921   No. 255

JOURNAL

OF

he New York Botanical Garden

EDITOR

R. S. WILLIAMS

Administrative Assistant

CONTENTS

PRICE $1.00 A YEAR; 10 CENTS A COPY

PUBLISHED FOR THE GARDEN
AT 8 WEST KING STREET, LANCASTER, PA.
INTELLIGENCER PRINTING COMPANY

PLATE 255

Mr. Billy Bowlegs in his corn-field on Indian Prairie, west of Lake Okeechobee. Bowlegs is a descendant of the famous Indian Chief of a century ago. His field was once a palmetto hammock as is evidenced in the plate, by the palm stumps still standing among the corn. The palmetto trunks have been made into a snake fence to be seen in the background, which separates the fields of father and son. The son's corn, several weeks older, may be seen on the far side of the fence. The soil of the prairie is sand, but in the palmetto hammocks the accumulated humus of ages makes it very fertile.

JOURNAL

OF

The New York Botanical Garden

| Vol. XXII | March, 1921 | No. 255 |

OLD TRAILS AND NEW DISCOVERIES

WITH PLATES 255 AND 256
(*Continued*)

FLORIDA KEYS

Two days later we set out for another cactus center, namely, the Florida Keys. The "Barbee" had started for Key West on Monday morning, and the writer left Miami the following day for that port by train. Both boat and train reached Key West at six o'clock Tuesday evening. Several among many interesting observations made from the train may be worthy of notice. First, it is often said that the cocoanut tree always grows with its swollen trunk-base curved towards the prevailing wind or toward the water. This does not hold in the case of the cocoanuts on the keys, for they grow in various postures. Second, vegetation has sprung up and now thrives on the lime-rock ballast of the concrete viaducts connecting different keys: beggar-ticks (*Bidens leucantha*), sand-spurs (*Cenchrus carolinianus, C. echinatus*), rush-grass (*Sporobolus domingensis*), horse-weed (*Leptilon canadense*), pepper-grass (*Lepidium virginicum*), and spurge (*Chamaesyce* sp.). These appeared to be the most abundant kinds. Third, a great many century-plants (*Agave*), both of native kinds and introduced, were in flower, a feature which gave, especially to Key Vaca and its neighboring islands, a most unusual aspect.

Cactus-hunting interested Mr. Cuthbert so much, that according to prearranged plans he joined us at Key West in order to make the cruise through the Keys to Miami.

We spent a day searching all parts of Key West for cacti. Prickly-pears were abundant, of course, and we found plenty of the recently described *Opuntia zebrina*, which was collected once before on that island, and also some scattered remnants of the tree-cactus (*Cephalocereus keyensis*) towards the eastern end of the key; but the main object of our search, a prickly-apple (*Harrisia*), failed to appear anywhere. After gathering some miscellaneous specimens we gave up our search there and prepared to leave the island early the next morning.

Big Pine Key was our next destination. A strong east wind prevented us from taking the outside or more direct course. Consequently, we started on the inside course, which runs in a somewhat semicircular curve around a labyrinth of small islands on the inner side of the reef. After bucking a bad sea for several hours, we located a stake about a mile east of Harbor Key and found our way into Big Spanish Channel, whence we picked our way gradually southward between No Name Key and Big Pine Key.

Inquiries at the settlement on Big Pine for the locations of hammocks with which we were not already acquainted failed to bring much encouraging information. Most of the former large hammocks had long ago been destroyed in making charcoal for supplying Key West, and furthermore, during our former explorations there we had never met with any of the kinds of cacti we now sought. However, the report of some hammock growth towards the upper end of the island decided us to retrace our course for about five miles and land at a point indicated by our informant. We were not long in locating the site of a former hammock by the character of the shrubs and an occasional surviving tree. An extended search in that region failed to disclose any cacti, however, except prickly-pears. We did find, though, a shrub new for Florida. This was a kind of candle-wood, *Dodonaea Ehrenbergii*, a plant first found on Hispaniola, and not previously known this side of the Gulf Stream.

Failing of our object there, we retraced our course to the settlement, where we made a search of several remnants of hammocks, but the search was wholly barren of results so far as cacti were concerned.

However, at last we heard of a hammock near the southeastern tip of the key which, so they said, hunters shunned "because cacti grew so thickly there that dogs couldn't get through." We at once returned to the "Barbee" and moved to the lower end of the island. After landing on the extensive plain that stretches beyond the pine woods, we located a charcoal-burner's trail and followed it. This trail led toward an insignificant-looking hammock, which, from a distance, the writer had theretofore taken to be a mere button-wood swamp. We entered the hammock and there, to our great surprise and delight, our eyes were confronted with cacti of no less than eight different kinds. There were not less than five genera.

But most pleasing to behold was that plant we particularly sought, a species of *Harrisia*.

Not many years ago only six kinds of cacti were recorded from all Florida, or, in fact, from all the Southern States; but here we had found eight quite distinct kinds growing in one small area. These cacti fall naturally and equally into two categories: *Opuntia Dillenii*, *O. keyensis*, *O. austrina* (maritime form), and *Acanthocereus pentagonus*, these representing common plants in tropical and subtropical Florida; *Cephalocereus keyensis*, *C. Deeringii*, *Harrisia* sp.?, and *Opuntia* sp.?, representing rare plants.

In order to understand the relationship of the *Harrisia*, however, we needed the flowers and fruits. Curiously enough, the numerous plants at the time of our visit had only young flower-buds and immature fruits. The prickly-pear just referred to had neither flowers nor fruits, but its vegetative characters showed it to be a new species of a group of the genus *Opuntia*, the Semaphores, none of which had previously been found this side of the Gulf Stream.

This most remarkable natural cactus-garden east of the western American deserts has maintained itself in spite of the fact that its other shrubby and arboreous elements have been cut out many times for furnishing wood for charcoal, charcoal burning having been in progress there for perhaps half a century. This is doubtless the hammock where Dr. J. L. Blodgett[1] and others in the earlier half of the last century collected

---

[1] John Loomis Blodgett (1809–1853) was born at South Amherst, Massachusetts, and studied medicine at Pittsfield. After a brief residence in the gulf states, and two years in Liberia, he settled at Key West, where he spent nearly fifteen years as physician and druggist. He was the first to make important botanical collections on the lower Florida keys.—JOHN HENDLEY BARNHART.

several typically West Indian shrubs that have not been found in Florida since. In fact, our incursion into that hammock was perhaps the first of its kind in more than three-quarters of a century. It is hoped to make an early survey of that locality for plants other than cacti.

This locality proved extremely interesting, not only from a botanical standpoint, but also from that of geology. Although it has long been known that the main groups of the Florida Keys, the Lower and the Upper, were in a general way separated by the Bahia Honda Channel, the writer has never been able to get information as to the real line of demarcation. The position of this hammock on the southern tip of Big Pine Key, however, gave a clue, at least, to the real line of division between the groups. The hook at the southern end of Big Pine Key has always seemed an extraneous part of the island. It is largely covered with marl, but the examination of this hammock showed that the rock structure is coral and not oölitic limestone. It is thus of a different geological period from that of the body of Big Pine Key. It is coral built up on the lower edge of Big Pine and really formed of an island now a peninsula, with an east-west long axis similar to that of the Upper Keys. Now to prophesy a little: Newfound Harbor Keys and Loggerhead Key also, with an east-west long axis, will doubtless be found to be of coral rock, instead of limestone, as they are now indicated on geological maps. The determination of this point and an exploration of their hammocks are looked forward to with much interest. If this prophecy comes true, the line of demarcation between the two groups of Keys will not be a north-south line, running between Bahia Honda Key on the east and Little Pine Key and No Name Key on the west, but an east-west line, from a point between Bahia Honda Key on the south and No Name Key on the north, running across the isthmus south of the limestone or pineclad part of Big Pine Key on the south, and the limestone keys on the north. Thus, instead of a complete separation of the two groups of islands, we would have an overlapping of the groups where they approach each other.[1]

[1] Since this paper was in type, the writer found that the conditions existing at the southern end of Big Pine Key and on one of the Newfound Harbor Keys, as stated above, were recorded by Samuel Sanford in a paper on "The Topography and Geology of Southern Florida," published in the Second Annual Report of the Florida State Geological Survey.

Among the interesting phenomena at this meeting point of the coral-rock and the oölitic limestone—and they are too numerous to record in this paper—may be mentioned the intermingling of the two tree-cacti of the East. In spite of the devastation of half a century, we found many fine trees of the key tree-cactus (*Cephalocereus keyensis*) formerly known only from Key West, and reported from Boca Chica Key, and the Deering tree-cactus (*Cephalocereus Deeringii*), heretofore known only from coral-rock on Umbrella Key and the Matecumbe Keys,[1] all indicating an interesting meeting place of plant species as well as of geological formations.

Having secured the desired specimens, we set out in the evening for Madeira Bay. We ran over the reef outside as far as the Bahia Honda Channel, where we turned into the Bay of Florida and started up the inside channel. A short stop was made on Bahia Honda Key. This differs from most, or perhaps all of the larger keys in having the rock foundation covered with a rather thick accumulation of sand; it is really a large but rather low sand-dune. On it two plants may be found in abundance that are not known to grow elsewhere this side of the Gulf Stream. One of these is a copiously branched and very copiously flowered vine of the morning-glory family (*Jacquemontia jamaicensis*) which has never been found elsewhere on the Florida Keys. The other is a remarkably spine-armed shrub of the madder family, *Catesbaea parviflora*, a relative of, but more diminutive in every way than the lily-thorn or spine-apple (*Catesbaea spinosa*) of the Bahama Islands, where our plant is also a native. It formerly grew on Big Pine Key, but it is apparently extinct there now. There is no tall vegetation on the island, and besides the herbaceous elements and scattered, scrubby, woody plants, low palms are usually the most conspicuous objects in the landscape.

After getting under way again, the east wind freshened and we decided to go as far as Bamboo Key and anchor for the night in its lee. Night fell before we reached Key Vaca, but we continued steadily and rapidly towards our objective, evidently more rapidly than we suspected, for the island we took to be Bamboo Key turned out to be Channel Key, at least six miles

[1] Journal of the New York Botanical Garden **17**: 198. 1916; **18**: 199-203. 1917.

beyond Bamboo Key; for unsuspectingly, in attempting to run in behind the island, we piled up on a shallow sand-bar, nothing the like of which is about Bamboo Key. By reversing the engine at once we pulled the boat off without delay. Bamboo Key is so low and devoid of hammock[1] that we had evidently passed it unobserved in the dark.

We had made better time than we suspected, owing doubtless to the increased power of the motor, as a result of the damper air of the night; but our Bahamian crew insisted that it was due to the fact that "water is always thinner at night than in the day time!"

Having passed our proposed anchorage, we decided to keep going until we reached Long Key, about five miles distant. By successfully locating the stakes of a channel through some shallow banks and keeping well to the north of some wreckage with which we were acquainted, we soon reached the anchorage.

At sunrise we resumed our journey and called at Lignum Vitae Key. There we secured a supply of a peculiar kind of prickly-pear for study, and then made all possible speed towards Madeira Hammock. During the preceding evening thunder storms were around us in all directions, but the weather cleared and we were treated to a wonderful sunset. The atmosphere during the following day was exceptionally clear and extremely bright. It was impossible to locate the horizon, and small keys in the distance seemed to stand high above the water, thus looking not like islands but aeroplanes.

As soon as we cast anchor at the entrance to Madeira Bay, we crossed the bay in a row-boat and went ashore, where we found more specimens of *Harrisia*. The plants of this genus are either terrestrial, or epiphytic, and sometimes grow on the trunks or branches of trees ten or twelve feet above the ground.

Madeira Hammock is full of animal life, and we invariably have experiences there with some branch of the animal kingdom—sometimes with rattlesnakes, often with yellow-jackets, oftener yet with mosquitoes. On this occasion one member of our party was enabled to study wasps at close range. He collided with a nest. In addition, the horse-fly season was on in southern Florida. Although we had met with only a few flies

[1] Journal of the New York Botanical Garden 18: 107. 1917.

in the hammock, when we returned to the "Barbee" the cabin was swarming with them, and we were well on our way before we were rid of the pests.

The sun was setting as we sailed away from Madeira Bay. After passing Pigeon Key we spent the evening hunting or rather haunting bars, banks, and barriers. If we did not at once find the channel stakes on a bank we would run along the bank in one direction or another until we picked them up. Thus, after negotiating one channel after another, we came to anchor at midnight in Black Water Sound, about a mile west of Snake Point and Jewfish Creek. With the wind in our favor we hoped to get a few hours of uninterrupted sleep safe from the attack of mosquitoes. Our hopes were realized—for two hours. Then all on board awoke at the same time, fighting for life. The mosquitoes had "boarded" us in a cloud. The battle lasted an hour. Then, suddenly, the mosquitoes had disappeared, and all hands fell off to sleep again as peacefully as if nothing had happened and slept soundly until awakened by dawn. Whence those mosquitoes came and whither they went is still a mystery.

The sun rising on Barnes Sound found us on the last leg of our cruise, which terminated in the early afternoon at Miami, where we disembarked and cared for the collections.

Miami and Marco Island

After an interval of three days, which was devoted mainly to local investigations in the vicinity of Miami and in cactus studies at Buena Vista, a third extensive excursion, again by land, was inaugurated. The country from the mouth of the Caloosahatchee to the Ten Thousand Islands was our most distant field. Our route was the same as that of the excursion to the Manatee region as far as Okeechobee City. Two weeks had elapsed and the season had advanced.

The two conspicuous and strongly contrasted plants of the "scrub" along the eastern coast were the Caribbean-pine (*Pinus caribaea*) with dark-green foliage, and the saw-palmetto (*Serenoa serrulata*) with whitish foliage. The leaves of the saw-palmetto that grows in the "scrub" are nearly always glaucous or grayish-white. In the rocky pinelands of the Everglade Keys the leaves are either green or grayish-white, while in the sandy pine woods and hammocks in other parts of the state they are typically green. The palm plants of the "scrub" are typically more robust than those of other localities. The reason for these variations presents a problem for some one to solve.

The beautiful, not to say elegant, tar-flower or fly-catcher (*Befaria racemosa*) stood in noticeable clumps here and there in the pine woods and high prairies. The low prairies, both in the coastal region and inland, seemed covered with drifts of golden-yellow snow, if such a phenomenon can be imagined, on account of the copious growth, bearing countless myriads of flowers, of bladder-worts, mainly *Stomoisia*, and the numerous ponds were thus made into mirrors with golden-yellow frames. In some of the lower prairies where the bladder-worts were less plentiful, other colored flowers predominated; meadow-beauties (*Rhexia*) and orchids were often abundant. Among the latter, grass-pinks (*Limodorum*) furnished bright colors, while the dainty-orchis (*Gymnadeniopsis nivea*), though lacking color, was conspicuous on account of its habit, suggesting so many tiny lamps scattered over the prairies.

If it were not for these numerous showy plants, Hungry Land would be desolate indeed. It is practically uninhabited, and in addition to the moss-draped hammocks and cypress-swamps, in some places the Florida-moss (*Dendropogon usneoides*) hangs in long streamers from the pine trees. Even Indians do not live there.

The land on the western side of Hungry Land slough, the Allapattah Flats, is somewhat less depressed and less depressing. It is inhabited by few white people, but by the Seminoles in considerable numbers, as we discovered by experience. Since our recent journey through that land, heavy rains had submerged parts of the trail. At one submerged point where the trail forked we inadvertently took the left-hand branch instead of the right and in a short time found ourselves in the midst of a number of Indian camps. We soon retraced our mistaken course and hastened on through the Allapattah (Seminole for alligator) Flats.

Plant life was represented in many interesting ways. However, two phenomena were particularly impressed on the writer: the often extensive turf formed of a sedge (*Eleocharis*) and the brilliant yellow of a Saint John's-wort (*Hypericum aspala-thoides*), which, when the numerous flowers were in large masses, presented such a striking yellow that it was almost painful to the eyes. Steady traveling soon brought us to Okeechobee City, whence we set out westward for the crossing of the Kissim-mee River. The vegetation changes suddenly beyond the pine woods west of Okeechobee City. Palmetto hammocks appear, as do also live-oak hammocks, both dense and open, and ham-mocks of mixed trees. On the prairies a maze of cattle trails run between and among myriads of large broom-like clumps of a stiff grass (*Spartina Bakeri*) and peculiar circular patches of the saw-palmetto. The latter plant was then in flower nearly everywhere and the violet-scented fragrance of its flowers filled the air.

After crossing the Kissimmee at Cabbage Bluff we headed southwest for the Caloosahatchee. There were no made roads as far as Fisheating Creek, so we were able to make good speed. The natural packing of the sand of those prairies is so firm that the traffic of horses, oxen, wagons, and, of late, automobiles, over a trail never seems to impair it to any serious extent; but when it is disturbed so as to make a roadbed, it soon be-comes practically impassable, until it is surfaced with some hard material.

Out across the great palmetto prairie we presently pene-trated probably the most remarkable growth of cabbage-trees (*Sabal Palmetto*) in existence. This palm grows there singly or in groups of dozens, hundreds, and thousands, forming groves essentially to the exclusion of all other trees. · The magnitude of the growth was most impressive and often most beautiful.

The trail winds in and out among the hammocks in blind curves, and as we took one of these we nearly collided with an Indian, Billy Stuart, who was out with his family on a "joy-ride" in his ox-car. As he was making perhaps one mile an hour we averted a head-on collision. We even stopped for a short visit and before we separated Billy invited us to call at his camp, at the same time explaining to us how to find it by leaving the trail at a designated hammock, on our return trip.

By the time we reached Fisheating Creek it was dark, and cloudy as well. But we crossed the creek and found the trail leading to the Caloosahatchee. When about half-way to the river we lost the trail, though, and later when the clouds cleared the stars indicated that we were traveling through the open pine woods towards the southeast instead of on the prairie to the southwest. Several lights appeared far ahead, so we hastened toward a supposed settlement to inquire concerning our whereabouts. When we came near the lights, instead of houses we found some burning stumps, the remains of yet another forest fire. Consequently, the only alternative was to double our course until we recovered the trail. This done, we made the Caloosahatchee in good time, crossed it at LaBelle and followed it to Fort Myers, reaching our destination just before midnight. We had covered the distance from Miami in about fifteen hours, including time for lunch and that consumed in twice losing the trail.

The next morning we set out for our most distant objective, namely Marco and vicinity. Traveling southward, after passing through several miles of sandy pineland, we came to the Lossmans River limestone which supports a succession of cypress swamps, and which was in some places overlaid with ancient sand-dune, the "scrub." The cypress trees of the swamps were sometimes associated with a peculiar-looking pine, the trunks of which were very slender and tall. After crossing several streams, whose bordering hammocks would also doubtless prove to be interesting collecting grounds, we came to the settlement of Naples, a town which is built on a spur of the "scrub," that here abuts directly on the Gulf of Mexico. Here, moreover, we found the northern limit of the Ten Thousand Islands, a region which extends southward along the coast to near the Northwest Cape of Cape Sable, or a distance, in a direct line, of about seventy-five miles. We crossed areas of "scrub," back of the coastal islands, about half way between Naples and Marco, which region perhaps represents the southern known limit of the "scrub" on the western side of Florida. The "scrub" thus extends about equally as far south on the west coast as on the east.

The higher land there rapidly fell away and gave place to extensive salt marshes or prairies, where the more conspicuous vegetation was made up of the cabbage-tree (*Sabal Palmetto*), the saw-palmetto (*Serenoa serrulata*), and buttonwood (*Conocarpus erecta*), all of which, or particularly the two palms, thrive equally well under the influence of salt water or fresh water, and also in either wet land or dry. Finally, we came within sight of the settlement of Marco, fifty-odd miles south of Fort Myers, where the white race is now engaged in the same pursuit that its prehistoric red race followed, namely, the oyster and clam business.

After arousing some of the inhabitants sufficiently to encourage them to cross the sound and learn our desires, we found that, with the usual foresight of such people for business, one of the ferry-lighters was at the bottom of the sound and the other was on the ways being repaired. Thus, the main object of our excursion, an examination of the hammocks about Caxambas Pass, was defeated. Leaving our car on the northern shore, we crossed the sound for a short investigation of the flora of the immediate vicinity of the settlement of Marco, leaving the Caxambas end of the island for future attention. This completed, we recrossed the sound and began to retrace our course northward.

Hard luck had evidently overtaken us at last.

An anticipated visit to the Royal Palm Hammocks of the Cape Romano district was also defeated, although we managed to get within about four miles of our objective. After repeatedly "bogging" in the dry sand of a side trail on the one hand and in the soft mud of the adjacent prairies on the other, we sought the main trail and set out for Fort Myers. Herbarium specimens were collected from the different plant associations. But the "scrub" was negatively interesting, that is, the season seemed to be very backward and very few plants were in flower. However, one notable shrub, a kind of lead-plant, an undescribed species of *Amorpha*, was found in full bloom. This plant was quite conspicuous, both in the matter of foliage and inflorescence, and, curiously enough, it seems to be most closely related to the lead-plant (*Amorpha canescens*) of the western plains.

The following morning we drove to Punta Rassa, which is situated at the mouth of the Caloosahatchee. With the co-operation of several fishermen, we were able to visit a number of islands at the mouth of the river. At Punta Blanca, at the mouth of the Caloosahatchee, the most likely place for the occurrence of *Harrisia*, we failed in our search, although the plant was said to grow there formerly. Many of the plants usually associated with *Harrisia* were present, and several other kinds of cacti were plentiful. However, the locality proved very interesting historically as well as botanically. Unfortunately, most of the hammock had long been destroyed by the digging away of the shells of the once evidently important kitchen-midden. The shells and numerous human remains had been and are still being carried away for making roads in the vicinity of Fort Myers.

This locality, like the shores of other estuaries, was probably one of the places of rendezvous for the fierce Calusas, the aborigines of southern peninsular Florida and of the Keys. It was in this vicinity, it is said, that the Calusas captured and held as prisoners for a century, at least, an expedition of Caribs from the West Indies in search of a fountain of youth. They were maintained as a separate settlement. It may be that the Big Cypress Seminoles (Creeks) have the blood of the old Calusas and Caribs in their veins.

The most interesting object at Punta Blanca was the old fire-place and perhaps also sacrificial altar of the Calusas, for human sacrifice is said to have been practiced. The constant fires burned the accumulated shells into a solid mass of lime, which increased in height as the kitchen-midden was gradually built higher. When the shells in recent times were removed for road-making material, this concreted fire-place was spared. But to pass from more ancient history to more modern: One of our fisherman navigators related to us the local interpretation and belief regarding the occurrence of the human skeletons in the shell heaps at Punta Blanca. The story runs as follows: In the early part of the last century a vessel bringing a large number of negro slaves from Africa was discovered and chased by a fleet United States revenue cutter. The slave vessel took refuge in the waters about what is now called Punta Blanca when the revenue cutter opened fire and killed all the negroes. Their bodies were then buried on the adjacent shore!

The same evening found us speeding up the valley of the Caloosahatchee. We reached a frontier settlement about bedtime and spent the night there, a night that will long be remembered, for giant cockroaches, big spiders, and scorpions played hide-and-seek over the bed till daylight. Dawn was welcome

PLATE 256

In a slough north of Eagle Bay, Lake Okeechobee. This slough, now filled with a pop-ash hammock, was evidently, when the land was less elevated, the bed of a river which flowed from the prairies into Lake Okeechobee. Even now, water stands in it to the depth of one to three feet during part of the year. The ash trees consequently are amphibious. Notice the peculiar growth of the trees. Several stems or branches arise from a large, partly bruised base. The numerous epiphytes on the trees are several kinds of wild-pines (*Tillandsia*) and an orchid (*Encyclia tampense*).

when it at last arrived and we made haste to get away over the prairie trail. The massive hammocks on Fisheating Creek and its tributaries and adjacent sloughs were not only conspicuous, but beautiful. The pale under-sides of the sweet-bay leaves turned up by the breeze showed in striking contrast against the deep green fresh foliage of the associated trees.

The prairie flowers became more numerous day by day. Spider-lilies (*Hymenocallis*) were scattered throughout the turf. Marsh-pinks (*Sabbatia*) and meadow-beauties (*Rhexia*) formed bright patches in all the slightly depressed places. But the flower most notable, not on account of its size, but because of its countless numbers, was the yellow-eyed grass (*Xyris*) whose heads stood above the other herbaceous vegetation nearly everywhere as far as the eye could see.

In order to fulfill our social duty and make that call at Billy Stuart's camp, we left the trail where directed to do so. We had not proceeded an eighth of a mile before we saw an Indian perched on a stump about a quarter of a mile away, evidently with his eye on us. We met him and found he was Billy Bowlegs, evidently a descendant of the former celebrated chief of that name. He directed us to Billy Stuart's camp, which we found without difficulty. There were many camps in that region, all situated in palmetto hammocks. The hammocks, when cleared of the palms and shrubbery, also furnished the fields for growing corn, and many fine crops of maize were observed. We had a long talk with Billy Bowlegs in his cornfield. Among other things he said he liked cultivating crops very well; but hunting was more to his liking.

The social functions being over, we made all possible haste to cross the Kissimmee River. A boy who was on the ferry directed us to a short trail running from the Kissimmee to Okeechobee City. We could hardly believe him, but we agreed to try the new course and followed his direction. We found he was right and were surprised to learn that nearly half the distance of the trail we were used to traversing was eliminated. But, more interesting than the saving of distance was the finding of a new and very characteristic type of hammock. It was only a few hundred yards wide, but an indefinite number of miles long. The arboreous growth was almost exclusively of an ash (*Fraxinus*), but its habit was different from that of any ash with which the writer is acquainted. The very short and broad bases of the trees were divided usually into from a half dozen to two dozen trunks. The trunks and limbs of the trees were copiously laden with epiphytes, both orchids and bromeliads, the latter predominating. The conspicuous orchid was the tree-orchid (*Encyclia tampense*), the only epiphytic orchid of southern Florida that does not definitely occur also in the West Indies. The bromeliads were chiefly *Tillandsia recurvata*, *T. bracteata*, *T. utriculata*, and the long-moss, *Dendropogon usneoides*.

We had so far been favored with good weather, but were now completely surrounded with severe thunder storms and we made haste to get into the region east of Lake Okeechobee. Along the recently opened right-of-way between Okeechobee City and the lake we found a sunflower (*Helianthus cucumerifolius*), a native of eastern Texas, naturalized in the ditches. A thistle (*Cirsium*), discovered on the eastern shore of the lake several years ago, was very plentiful in the cleared hammock land. This thistle has numerous close-set leaves and many medium-sized flower-heads, and it grows to a height of twelve feet. The discovery of the cucumber-leaved sunflower, as a naturalized plant, was fortunate, as it gave the writer an opportunity to compare it with its Florida relative, *Helianthus debilis*. He has maintained these very distinct species, although the "closet botanists" have for over a century considered them to represent but one species.

The wide beach, lately the bottom of the lake, was covered in most places with a sedge (*Cyperus LeContei*) and dog-fennel (*Eupatorium capillifolium*). The latter plant, as well as other tall herbs, were veiled with dense white spider-webs which were drawn tightly around the foliage. Blown to and fro by the wind of an approaching storm these veiled plants appeared like thousands of shrouded ghosts moving over the wide shore. By collecting quantities of these spider-webs on their horns the wild cattle roaming on the lake-shores all have the appearance of wearing white nightcaps.

The now nearly spent storm, as well as night, overtook us just after we passed through the Okeechobee hammock and light rain accompanied us through the Allapattah Flats. Hungry Land had evidently had a drenching rainstorm, as nearly all the land, together with the numerous plants mentioned on a preceding page, was submerged. Thus flood and darkness put a stop to further collecting.

Minor investigations were prosecuted when we were not absent from Miami on these extended excursions. Among several localities of interest visited, the region back of Cape Sable may be recorded. We passed through Royal Palm Hammock, where it was gratifying to see the several acres of former forest that was fire-swept and completely destroyed a few years ago, rapidly reforesting itself. Since the custodian of the park cleared the debris from the devastated area within the last two years, the progress of the natural growth has been remarkable. Under Mr. Mosier's guidance we visited a hammock about twenty miles southwest of Royal Palm Hammock. This hammock is somewhat similar to another in the same region, mentioned in a former paper.[1] As we gain more knowledge of the Long Key Pineland and Cape Sable it appears that a disconnected chain of pine-islands and high hammocks connects the two regions, and the intervening territory, when accessible, will doubtless prove an extremely interesting collecting ground.

Our visit to the hammock mentioned above was mainly for the purpose of collecting a peculiar epiphytic fern that had been found there several days earlier. The fern seems to be a species of tropical polypody (*Polypodium Plumula*) with much elongated leaves. These were up to about three feet in length, while they were only between one and two inches wide. The blades were tightly curled, somewhat after the manner of the resurrection-fern, when we found them, and three days' time was required to expand them after they were placed in water.

The spring vegetation of the Everglades was well advanced and the difference between the high and low grounds of the prairie was there more pronounced than had been noticed elsewhere. The lower areas are really natural drainage sloughs; but they are in strong contrast with the deeper sloughs. The latter, in which humus, as well as moisture, collects abundantly support a copious growth of herbs and often also of shrubs, while the former, being shallow and quite even, retain neither humus nor moisture and in addition are subject to frequent

[1] Journal of the New York Botanical Garden **20**: 191–207. 1919.

prairie fires. These are barren or nearly barren in the dry season, and thus stand out in strong contrast with the higher and well plant-clothed parts of the prairies.

The Everglades presented quite a different appearance from what they did six months previously. This was particularly so in the regions where the cypress grew. The distant massive cypress heads instead of presenting a mass of gray branches, showed up as immense green domes, while the stunted, scattered, or spaced pond-cypress on the prairies had been transformed from scraggly spectres[1] into trees with light green foliage and copious tassels of flowers.

Different also was the review of the woods and fields as we sped northward on our return. Summer was there and everywhere, whatever the calendar may have said.

JOHN K. SMALL.

SUCCESSION OF ALGAE IN THE GRASSY SPRAIN RESERVOIR[2]

Grassy Sprain Reservoir, the Yonkers City reservoir, is a long narrow lake made by damming a small valley. At the southern end is a spillway with a small stream flowing down over a series of steps. At the bottom a channel extends back obliquely for a hundred yards to the dam, this channel is filled with an almost stagnant backwater. Where the backwater and the overflow stream meet is a spring with a barrel set in it. The water in the spring and in the outlet was open all of the winter of 1919–1920, but the reservoir was frozen over solidly. Efforts were made to collect algae by cutting through the ice but without success. Algae were collected in the spring during February and March and in the spring, backwater and along the south and west sides of the reservoir, twice a month during the remainder of the year.

No attempt was made to classify the numerous diatoms. Numerous desmids were determined but are not considered in the present report. The filamentous green algae were abundant from the time the ice melted in March through July, the largest amounts being found in May and June when large floating

[1] Journal of the New York Botanical Garden 21: 53. 1920.
[2] Account of report before the Garden conference January 5.

masses of *Mougeotia*, *Zygnema* and *Spirogyra* were common all along the shore. From August through November few green algae were found, but the number began to increase early in December.

Many of the algae had a very short period of growth. For example, *Spirogyra crassa* first appeared early in June at the end of a bay in large masses lying on the bottom. A week later it was still more abundant, the lustrous brilliant green masses covering over half the area of the bay and a few small masses were floating and had been carried out from the bay quite a distance. It continued abundant till near the end of July, when only a little was to be found and that sparingly in fruit. In the first week of August no trace of it could be found.

For six months, from February till July, in the backwater of the outlet there were large masses of *Spirogyra Weberi*, at times filling the channel from side to side. During June these masses became smaller and yellowish in color and by the end of July all had disappeared. In October small masses similar to those found there in the early spring appeared but proved to be *S. fluviatilis*. By January this had increased till it was as abundant as *S. Weberi* had been in March, but no trace of the latter was to be found.

In February the edge of the barrel in the spring of the outlet was fringed with a bright green mass of *Stigeoclonium subsecundum*. It increased in length during the following two months, forming streamers several centimeters in length. In May and June the amount decreased and the plants found showed signs of disintegrating. From the end of June till December none was found except a few short filaments mixed with *Oscillatoria* on the inside of the barrel. In December it again appeared on the edge of the barrel, and by January was as abundant as during the previous February. Whether small particles of the plants had remained attached to the wood or the rhizoid-like holdfasts remained in the pores of the wood, or whether spores of the plant carried it over was not determined, but it seems probable that part of the plant remained on or in the grain of the wood.

Of the blue-green algae some were found every time collections were made. The more or less spherical masses of *Plectonema tomasianum* attached to plants below water or floating free on the surface were especially abundant in July and August but apparently are present all through the year.

In the early part of August the bottom of the lake was dotted everywhere in shallow water with dark green patches of *Aphanocapsa Grevillei* from which slender threads or streamers, each with an enclosed gas bubble at the upper end, rose to or towards the surface. In the latter part of September similar patches, without the streamers, were composed chiefly of *Cylindrospermum catenatum*, a little of which had formerly been mixed with the *Aphanocapsa*.

About the last of September the lake "bloomed," being covered with dots of *Clathrocystis aeruginosa* so closely as to make the water cloudy. A week later the amount of the little masses was so great that they piled up along the shore, due to wind action, making piles from one-fourth to an inch thick, of the consistency of soft putty and with a disagreeable seaweed odor.

The blue-greens reached their maximum both in numbers and abundance during September. The green algae were most abundant in late May and early June and were in the smallest amount when the others were at their maximum.

G. T. HASTINGS.

NOTES, NEWS AND COMMENT

Three species of witch-hazel were in bloom in the fruticetum in February. The most decorative of these is *Hamamelis mollis*, a native of China; its flowers are much larger than in the others and are deliciously fragrant; for horticultural purposes it is much to be preferred. The Japanese witch-hazel, *H. japonica*, also in bloom at the same time, is interesting but its flowers are relatively inconspicuous. The third species, *H. vernalis*, is a native from Missouri and Oklahoma to Louisiana, and in its flowers and blooming time differs materially from the fourth species in the fruticetum, *H. virginiana*, the common witch-hazel of the eastern United States, which blossoms in the late fall and early winter. There is a fifth species, not in the fruticetum collection, recently discovered, *H. incarnata*, from Japan, which differs in having the petals a deep red and the flowers on long drooping stems. These are the five species known at present: two from Japan, one from China, and two from the United States.

Many of the orchids in flower in conservatory range 2 are assembled in the south end of house 2A. There has been an interesting display during the winter. Since January 1 about 150 species have come into bloom. There is a great diversity in the color, size, and form of the flowers and general appearance of orchids, but the essential character which distinguishes them from all others of the monocotyledons, the uniting into a column of the pistil and stamens, is present in all the flowers.

The large collection of aroids, normally located in house 10 of conservatory range 1, is temporarily placed in house 14 of the same range, pending a renovation of the former house. The center bench in house 14 has been removed, and the larger aroids are placed on the ground or are slightly elevated. In this position the beauty of the plants can be much better appreciated than when on a bench.

A plant of the chocolate tree, *Theobroma Cacoa*, is in fruit in conservatory range 1, house 3. There is a solitary fruit, but it is maturing normally. In this plant the flowers and fruit are borne on the trunk and the bases of the branches, and not, as in most other trees, on the ends of the branches. Each pod contains five rows of seeds, the total content being from fifty to seventy-five seeds, which, when dried, form the chocolate beans of commerce.

Dr. and Mrs. Britton, accompanied by Dr. F. J. Seaver, sailed for Trinidad on February 21, in order to continue the botanical exploration of that island. They expect to return about the first of May.

The following visiting botanists have registered in the library during the winter: Prof. Elmer D. Merrill, Manila, P. I.; Prof. P. J. Anderson, Amherst, Mass.; Prof. C. R. Orton, State College, Pa.; Mr. Charles C. Deam, Bluffton, Ind.; Prof. W. A. Setchell, Berkeley, Cal.; Mr. Ellsworth P. Killip, Mr. E. G. Arzberger, Mr. L. L. Harter and Dr. J. N. Rose, Washington, D. C.; Prof. Alfred H. W. Povah, Syracuse, N. Y.; Prof. W. W. Rowlee, Ithaca, N. Y.; Mr. A. B. Massey, Blacksburg, Va.;

Mr. G. W. Martin and class, New Brunswick, N. J.; Prof. H. M. Fitzpatrick, Cambridge, Mass.; and Prof. Caroline A. Black, New London, Conn.

Meteorology for February—The total precipitation for the month was 3.23 inches, of which 1.60 inches (16 inches by snow measurement) fell as snow. The maximum temperatures recorded for each week were as follows: 50° on the 6th, 46° on the 8th, *62° on the 16th*, and 42° on the 22nd, 23rd and 28th. The minimum temperatures were: 15° on the 1st, 25° on the 9th, 12° on the 21st, and *6° on the 25th.*

ACCESSIONS

PLANTS AND SEEDS

1 plant of *Iris* for conservatories. (By exchange with Royal Botanic Garden, Regents Park, London.)

1 plant of *Mamillaria erecta*. (By exchange with U. S. Nat. Museum, through Dr. J. N. Rose.)

7 plants derived from seeds from various sources.

1 plant of *Daphne laureola*. (Given by Mr. E. G. Pendleton.)

1 plant of *Thea japonica alba*. (Given by Mr. Jos. Nickel.)

2 plants of Nerium Oleander. (Given by Mrs. Arthur C. James.)

16 plants of *Iris*. (Given by Bobbink & Atkins.)

6 plants of *Iris*. (Given by Vaughans Seed Store.)

56 plants of *Iris*. (Given by the Wing Seed Co.)

2 plants of *Iris*. (Given by Mr. H. W. Groschner.)

13 plants of *Iris*. (Given by Sunnybrook Farm.)

3 plants of *Iris*. (Given by the Mount Desert Nurseries.)

124 orchid plants from Brazil. (Given by Mr. Lee G. Day.)

1 plant of *Iris*. (Given by the Good & Reese Co.)

5 plants of *Iris*. (Given by Mr. Carl Purdy.)

105 plants of *Iris*. (Given by the Van Wert Iris Garden.)

31 plants of *Iris*. (Given by Mr. Robert T. Jackson.)

3 plants of *Iris*. (Given by Mr. H. E. Eckert.)

10 plants of Cacti for conservatories. (By exchange with U. S. Nat. Museum, through Dr. J. N. Rose.)

3 plants for conservatories. (By exchange with Cambridge Botanical Garden, England.)

6 plants of Cacti, for conservatories. (By exchange with Royal Botanic Gardens, Kew.)

1 plant of *Musa Ensete*. (By exchange with Mr. V. Soar, through Dr. J. K. Small.)

86 plants derived from seeds from various sources.

1 packet of seed. (Given by Mr. C. J. Chamberlain.)

2 packets of Cuban seed. (Given by Mr. Mario Calvins.)

2 packets of seed. (Given by Mr. P. D. Barnhart, through Mr. A. T. Delara.)

44 packets of seed, from New Zealand. (Given by Miss E. Mabel Clark.)

1 packet of seed. (Given by Mr. Geo. J. Chryssicus.)

172 packets of seed. (Purchased.)

1 packet of Colombian seed. (Collected by Dr. H. H. Rusby and Dr. F. W. Pennell.)

4 packets of seed. (By exchange with Bureau of Plant Industry.)

116 packets of seed. (By exchange with Botanical Garden, La Mortola, Italy.)

77 packets of seed. (By exchange with Botanical Garden, Berne, Switzerland.)

84 packets of seed. (By exchange with Botanical Garden, Gothenburg, Sweden.)

71 packets of seed. (By exchange with Botanical Garden, Glasnevin, Dublin.)

24 packets of seed. (By exchange with Bureau of Plant Industry.)

118 packets of seed. (By exchange with Botanical Garden, Lyons, France.)

474 packets of seed. (By exchange with Botanical Garden, Edinburgh, Scotland.)

38 packets of seed. (By exchange with Botanical Garden, Upsala, Sweden.)

246 packets of seed. (By exchange with Botanical Garden, Zurich, Switzerland.)

83 packets of seed. (By exchange with Botanical Garden, Groningen, Holland.)

1 packet of seed of *Phormium tenax*. (Given by Mr. Jos. Dixon.)

1 packet of seed. (Given by Mr. H. W. Foulkner.)

2 packets of Trinidad seed. (Collected by Dr. N. L. Britton.)

1 packet of seed. (By exchange with Mr. T. D. A. Cockerell.)

4 packets of seed. (Collected by Dr. N. L. Britton.)

1 packet seed *Penstemon Helleri*. (Collected by Dr. F. W. Pennell.)

1 packet tree fern spores. (By exchange with U. S. Dept. Agric.)

1 packet seed. (Collected by Dr. F. W. Pennell.)

1 packet seed. (Given by Miss Margaret Barrow.)

1 packet seed of *Gentiana crinita*. (Given by Mr. Herbert Durand.)

1 packet of seed of black sesame. (Given by Mr. Wm. Beekley.)

1 packet *Commelina* seed. (Collected by Dr. F. W. Pennell.)

1 packet seed of *Sabal texana*. (By exchange with Mr. R. Runyon, through Dr. J. K. Small.)

MUSEUM AND HERBARIUM

50 specimens of flowering plants from New York. (By exchange with the University of the State of New York.)

38 specimens of drugs. (Given by Dr. H. H. Rusby.)

28; specimens of flowering plants from North America. (By exchange with the United States National Museum.)

6 specimens of woods. (Given by Dr. H. H. Rusby.)

2 specimens of *Limosella* from Colorado. (By exchange with Prof. Ellswor⁺ ı Bethel.)

11 specimens of fodder plants. (Given by Dr. H. H. Rusby.)

1 specimen of *Galium cruciatum* from New York. (Given by Mr. Wm. C. Ferguson.)

29 specimens of beverages. (Given by Dr. H. H. Rusby.)

6 lantern slides of dahlias. (By exchange with Mrs. C. H. Stout.)

8 specimens of volatile oils and perfumery. (Given by Dr. H. H. Rusby.)

2300 specimens of flowering plants from Trinidad and Tobago. (Acquired by Dr. N. L. Britton.)

7 specimens of spices. (Given by Dr. H. H. Rusby.)

8 specimens of fungi from California. (By exchange with Prof. Ellsworth Bethel.)

2 specimens of sugars. (Given by Dr. H. H. Rusby.)

250 specimens of flowering plants and ferns from northeastern Maine. (Collected by Prof. Harold St. John.)

14 specimens of proximate principles of plants. (Given by Dr. H. H. Rusby.)

161 specimens of flowering plants from Jamaica, West Indies. (By exchange with the Royal Gardens, Kew, England.)

33 specimens of foods. (Given by Dr. H. H. Rusby.)

100 specimens "North American Uredinales" fascicle 23. (Distributed by Mr. Elam Bartholomew.)

5 specimens of fibers. (Given by Dr. H. H. Rusby.)

36 specimens of varnish resins. (Given by Messrs. Pomeroy and Fischer.)

4 specimens of soap products. (Given by H. H. Rusby.)

8 specimens of flowering plants from Oregon. (Given by Professor J. C. Nelson.)

3 specimens of starches. (Given by Dr. H. H. Rusby.)

95 specimens of flowering plants from Montana and Idaho. (Given by Professor J. F. Kemp.)

21 specimens of poisonous plants and insecticides. (Given by Dr. H. H. Rusby.)

15 specimens of flowering plants from Montana, Oregon, and Washington. (Given by Mr. P. H. Hawkins.)

3 specimens of flowering plants from Colorado. (By exchange with Mr. I. W. Clokey.)

121 specimens of mosses from New Zealand. (By exchange with Mr. Brown.)

1 specimen of moss from France. (By exchange with M. Theriot.)

· 19 specimens from Iceland. (By exchange with Dr. Andrews.)

52 specimens of mosses from North Carolina. (By exchange with P. O. Schallert.)

67 specimens of mosses from California. (By exchange with Dr. C. F. Millspaugh.)

24 specimens of mosses from Minnesota. (By exchange with Prof. J. M. Holzinger.)

25 specimens of mosses from Florida. (Collected by Dr. and Mrs. N. L. Britton.)

1850 specimens of flowering and flowerless plants from Florida. (Collected by Dr. John K. Small.)

1 museum specimen of *Ilex vomitoria* from Florida. (Collected by Dr. J. K. Small.)

40 specimens of Tertiary fossil plants in clay, from Brazil. (Given by Prof. J. C. Branner.)

15 specimens of Tertiary (?) fossil plants in red shale from Trinidad. (Collected by Dr. N. L. Britton.)

2 specimens of shale showing the root-markings of plants in the cleavage planes. (Given by Mr. F. F. Burr.)

13 specimens of fungi from Washington. (By exchange with Prof. S. M. Zeller.).

7 specimens of fungi from Idaho. (By exchange with Dr. J. R. Weir.)

2 specimens of fungi from California. (By exchange with Prof. Ellsworth Bethel.)

1 specimen of *Lophodermium arundinaceum* from Indiana. (By exchange with Dr. H. S. Jackson.)

50 specimens of fungi Rehm, Ascomycetes exsiccati. (Distributed for Dr. H. Rehm.)

1 photograph of *Cordyceps herculea*. (Given by Dr. H. A. Kelly.)

1 specimen of *Calvatia collata* from Utah. (By exchange with Prof. A. O. Garrett.)

2 specimens of *M lanoleuca alboflavida* from Pennsylvania. (By exchange with Mr. H. L. Daddow.)

1 specimen of *Polyporus elegans* from California. (By exchange with Dr. E. P. Meinecke.)

1 specimen of *Psathyra* from the New York Botanical Garden. (Collected by Dr. W. A. Murrill.)

1 specimen of *Funalia stuppea* from Utah. (By exchange with Prof. A. O. Garrett.)

6 specimens of cup fungi from Michigan. (By exchange with Dr. H. M. Fitzpatrick & Geo. Hume Smith.)

5 specimens of cup fungi from New York. (By exchange with Prof. H. H. Whetzel.)

3 specimens of cup fungi from Idaho. (By exchange with Dr. J. R. Weir.)

2 specimens of fungi from local flora range. (Given by Mr. C. A. Schwarze.)

1 specimen of *Chantarel floccosus* from Harrison, Maine. (By exchange with Dr. G. Clyde Fisher.)

1 specimen of *Lachnocladium Schweinitzii* from Arkansas. (By exchange with Rev. H. E. Wheeler.)

1 specimen of *Pterula subulata* from Washington, D. C. (By exchange with Prof. H. H. Whetzel.)

15 specimens of fungi from Yama Farms, New York. (By exchange with Miss Grace O. Winter.)

1 specimen of *Melanoleuca melaleuca* from Arizona. (By exchange with Dr. Walter Hough.)

2 specimens of fungi from Utah. (By exchange with Prof. A. O. Garrett.)

40 specimens of fungi from Buck Hill Falls, Pennsylvania. (By exchange with Mrs. John R. Delafield.)

2 specimens of fungi from Pennsylvania. (By exchange with Dr. L. O. Overholts.)

2 specimens of *Russula compacta* from the New York Botanical Garden. (Collected by Dr. W. A. Murrill.)

25 specimens "Phycomycetes and Protomycetes," numbers 326–350. (Distributed by P. Sydow.)

100 specimens "Uredineen," numbers 2651–2751. (Distributed by P. Sydow.)

25 specimens "Fungi Dakotenses" fasicle 19. (Distributed by Dr. J. F. Brenckle.)

1 specimen each of *Pyropolyporus Bakeri* and *Merulius aureus* from Pennsylvania. (By exchange with Dr. L. O. Overholts.)

1 specimen of *Cycloporus Greenei* from Kentucky. (By exchange with Prof. Bruce Fink.)

1 specimen of *Lactaria atroviridis* from New Jersey. (By exchange with Dr. D. L. Millspaugh.)

1 specimen of *Fomes roseus* from Michigan. (By exchange with Dr. D. R. Sumstine.)

6 specimens of fungi from Wyoming. (By exchange with Mr. Simon Davis.)

1 specimen of *Lactaria theiogala* from New Jersey. (By exchange with Dr. D. L. Millspaugh.)

1 specimen of *Clavaria byssiseda* from North Carolina. (By exchange with Dr. W. C. Coker.)

2 specimens of fungi from California. (By exchange with Prof. H. E. Parks.)

2 species of fungi from Canada. (By exchange with Rev. Dr. Robert Campbell.)

1 specimen of *Marasmius* from the New York Botanical Garden. (Collected by Miss M. E. Eaton.)

10 specimens of fungi from New York. (By exchange with Mr. W. H. Ballon.)

8 specimens of fungi from Oregon. (By exchange with Prof. S. M. Zeller.)

1 specimen of *Wynnea americana* from Pennsylvania. (By exchange with Mr. C. E. Chardon.)

1 specimen of *Clavaria fusiformis* from Alaska. (By exchange with Dr. N. M. Cook.)

10 specimens of fungi from Jamaica. (By exchange with Mr. W. R. Maxon.)

1 specimen of *Clavaria* from Big Cottonwood, Utah. (By exchange with Miss Violet Barrows.)

33 specimens of fungi from Oregon. (By exchange with Mr. J. S. Boyce.)

1 specimen of *Lachnea scutellata* from Connecticut. (By exchange with Prof. A. H. Graves.)

Provisions for
Benefactors, Patrons, Fellows, Fellowship Members, Sustaining Members, Annual Members and Life Members

---

1. Benefactors

The contribution of $25,000.00 or more to the funds of the Garden by gift or by bequest shall entitle the contributor to be a benefactor of the Garden.

2. Patrons

The contribution of $5000.00 or more to the funds of the Garden by gift or by bequest shall entitle the contributor to be a patron of the Garden.

3. Fellows for Life

The contribution of $1000.00 or more to the funds of the Garden at any one time shall entitle the contributor to be a fellow for life of the Garden.

4. Fellowship Members

Fellowship members pay $100.00 or more annually and become fellows for life when their payments aggregate $1000.00.

5. Sustaining Members

Sustaining members pay from $25.00 to $100.00 annually and become fellows for life when their payments aggregate $1000.00.

6. Annual Members

Annual members pay an annual fee of $10.00.

All members are entitled to the following privileges:

1. Tickets to all lectures given under the auspices of the Board of Managers.
2. Invitations to all exhibitions given under the auspices of the Board of Managers.
3. A copy of all handbooks published by the Garden.
4. A copy of all annual reports and Bulletins.
5. A copy of the monthly Journal.
6. Privileges of the Board Room.

7. Life Members

Annual members may become Life Members by the payment of a fee of $250.00.

Information

Members are invited to ask any questions they desire to have answered on botanical or horticultural subjects. Docents will accompany any members through the grounds and buildings any week day, leaving Museum Building at 3 o'clock.

Form of Bequest

I hereby bequeath to the New York Botanical Garden incorporated under the Laws of New York, Chapter 285 of 1891, the sum of..........

Vol. XXII   April, 1921   No. 256

JOURNAL

OF

The New York Botanical Garden

EDITOR

R. S. WILLIAMS
Administrative Assistant

CONTENTS

PRICE $1.00 A YEAR; 10 CENTS A COPY

PUBLISHED FOR THE GARDEN
AT 8 WEST KING STREET, LANCASTER, PA.
INTELLIGENCER PRINTING COMPANY

JOURNAL

OF

The New York Botanical Garden

| VOL. XXII | April, 1921 | No. 256 |

THE LEAFY SPURGE BECOMING A PEST

About one hundred years ago, possibly even earlier, the Leafy Spurge, native of Europe, obtained a foothold in Essex County, Massachusetts; the actual date of its introduction is not definitely known, but a specimen preserved in the Torrey Herbarium of Columbia University, deposited at the New York Botanical Garden, was collected at Newbury, Massachusetts, by William Oakes in 1827. Oakes, who was a keen botanist, annotated the label of the plant sent to Dr. Torrey, "I want to find this at another place; have you met with it, or heard of it?" For many years it was known nowhere else in the United States, but in the first edition of Gray's Manual of Botany, published in 1848, the learned author remarked that it was likely to become a troublesome weed. John Robinson in his "Flora of Essex County, Massashusetts" (1880) records it as "very abundant near Newburyport on the line of the eastern R. R.;" this is probably Oakes' locality. In his "Cayuga Flora," published in 1886, Professor W. R. Dudley records the plant as found at Groton, New York, in 1875, remarking on its rarity. In the Catalogue of Plants of Michigan by C. T. Wheeler and Erwin F. Smith (1881), it is recorded as escaped from cultivation and infrequent in that state. In the sixth edition of his Manual, published in 1889, Dr. Gray added these western New York and Michigan records to its known distribution, remarking, however, that it was rare. In the first edition of Illustrated Flora (1897) I recorded its distribution as Massachusetts to New York and Michigan; by the time the second edition was published (1913) it had extended to Maine, Ontario and New Jersey; it is only in recent years, however,

73

that Dr. Gray's surmise of 1857 has become a fact, and farmers in Orange County, New York, and elsewhere now find this weed a menace to pastures; measures for its reduction or eradication are being taken, but the task is not an easy one, and the matter is of sufficient importance to have brought out editorial comment in the New York Herald on February 9, 1921.

Fig. 1. Leafy Spurge (*Tithymalus Esula* Hill).

The Leafy Spurge, also known as Faitour's Grass and Tithymal (*Tithymalus Esula* Hill, *Euphorbia Esula* Linnaeus), native of Europe, is a perennial herb, usually about one and one-half feet high, with narrow leaves about two inches long or less, those on the stem few, those on branches close together, the uppermost in a verticil subtending a several-rayed umbel of small flowers and large broad bracts. The plant has horizontal rootstocks from which it grows year after year, and these make its control as a weed very difficult.

The accompanying illustration will help to identify the species; if detected it should be forked or spaded out of the ground and burned, taking care to get all its underground parts.

While often regarded botanically as a species of *Euphorbia*, the plant is quite different from the tall and fleshy, cactus-like, true Euphorbias of Africa, a collection of which may be seen in our Conservatory Range 1, House 5.

N. L. Britton

TROPICAL TREES WITH LIGHT-WEIGHT WOOD.*

During the last three years the writer has had very favorable opportunities to study light-weight woods in the forests and jungles of Central America. The wood of species of *Ochroma*, known commercially by the Spanish name "Balsa," has, in the last decade, come into prominence as a material for purposes of buoyancy and insulation. *Ochroma* is exclusively a genus of the American tropics. It grows in the West Indies but not in Florida. It is found in southern Mexico and from these northern limits extends to northern Bolivia.

Ochroma is by no means, however, the only genus that contains species with light-weight wood. There are several other genera in the same family (Bombacaceae) with light wood, notably some species of *Ceiba*, and all, so far as known, of the species of *Cavanillesia*. However, some representatives of the family have relatively heavy, hard wood, as for example, *Bombax ellipticum* and species of *Pachira*. Light-weight woods are to be found in other families, as in the genus *Wercklea* (Malvaceae) *Heliocarpus* (Tiliaceae), *Cordia* (Borraginaceae) and *Erythrina* (Leguminosae). Perhaps the most remarkable of all is the wood of the Caricaceae and particularly of the genus *Jacaratia*, which is so soft that it would be more appropriate to call it a vegetable substance rather than a wood. Its stature,

* Abstract of address before the Conference of the Scientific Staff and Students of the Garden in February.

bark and growth in thickness by a true cambium layer, however, warrant its being considered a tree.

One of the many remarkable features of tropical vegetation is the fact that the heaviest and lightest woods are, locally, often immediately associated.

Light-weight woods fall into two categories: first, those where the elements are arranged in a relatively homogeneous mass, and second, those where soft parenchymatous masses alternate with more or less continuous bands of thick-walled fibrous tissue. The first is Homogeneous, the second Laminated wood, as distinguished in the following discussion. *Cochlospermum*, *Erythrina*, and in the most extreme degree, all the Caricaceae, have laminated woods. The nearest approach to this type of wood in northern latitudes is in the Juglandaceae, where as is well known, very thin bands of parenchyma occur in the annual rings. In many tropical trees the greater portion of the mass of wood is composed of soft parenchyma. On the other hand, homogeneous woods have no conspicuous segregation of parenchymatous masses and sclerenchymatous fibers. Typical of this are all our northern woods, with the possible exception mentioned above, and the soft woods of the tropics, *Ochroma*. *Wercklea*, *Bombax* and *Cordia*.

Three genera have been selected to illustrate types of light woods. *Ochroma* and *Cavanillesia* of the Bombacaceae and, *Jacaratia* of the Caricaceae, in which genera, so far as known, all the species have light-weight wood.

OCHROMA—In the United States the wood of this genus is best known by the name "Balsa," but by Spaniards and the aboriginals in the tropical countries, by many other names.

In the Tropical Rain Forest Region the balsa attains its largest size and greatest abundance. Trees a meter in diameter are frequently seen. The leaves of young trees are much larger and thinner than in old trees and the wood found in the juvenile state is noticeably lighter than that of the senile state. In the tropics the trees flower and fruit at the age of two to three years. The average growth in diameter under favorable conditions is rapid, not far from 5 inches per year during the juvenile state. The wood ranges in weight from 4 to 12 pounds per cubic foot. The heart wood is homogeneous, silvery-white to pink. The sap wood is always white. There is a semblance of annual rings frequently, but they are irregular in occurrence and are in all probability due to variations in climate which do not occur with much regularity.

CAVANILLESIA—In Panama these trees are called "Quipo" and in northwestern Colombia, "Macondo." There are two species in South America, one in the east and one in the west. The Brazilian tree is called "Barraguda," by the Portuguese and Bottle Tree by the English. (A picture of it is shown in Schimper's *Plant Geography* p. 361.) The Panama tree is *C. platanifolia*. It attains immense size. The branches and leaves are at the very top of the tree and the trunks simulate great columns. The strength of the trunks is principally in the thick fibrous bark. The wood is even lighter than Balsa. It is so fragile however, that it crushes more or less as the tree falls. It is homogeneous but a large percentage of the wood is parenchyma. The wood is so porous that air can readily be forced through it longitudinally. The large trees appear to be of great age, but our surmise is that they are rapid growers and by no means as old as they look.

JACARATIA—With the exception of one anomalous species in Mexico, this genus has heretofore been considered as confined to South America east of the Andes. In 1920 specimens were found in Panama and Costa Rica. The tree greatly resembles a *Ceiba* and may have been mistaken for it. The fruit is characteristic of the Papayas. The wood of the tree is very soft and spongy. After a portion of the bark was cut away a machete could be shoved to the center of the trunk with comparative ease. The wood is laminated. The parenchyma bands are very thick and constitute at least nine tenths of the mass of wood. The wood when fresh can be cut into blocks as one would cut up parsnips or turnips. Balsa and Quipo do not shrink excessively when dry, but the wood of Jacaratia when dried shrinks to å mere fraction of its original bulk. It is said that the wood of the South American trees is poisonous when eaten.

There appears to be certain concomitant features attending trees with light wood somewhat as follows:

1. Geographically they are confined principally to the Tropical rain forest region and to the belt of equatorial calms.

2. They are plants with a very extensive leaf surface.
3. Thick, fibrous bark characterizes most of them.
4. Their wood is white or at least very light in color.

W. W. ROWLEE

PUBLICATIONS OF THE STAFF, SCHOLARS AND STUDENTS OF THE NEW YORK BOTANICAL GARDEN DURING THE YEAR 1920

Barnhart, J. H. (Biographical notes) in Small, J. K., Of grottoes and ancient dunes. Jour. N. Y. Bot. Gard. **21**: 30–36. Issue for Mr 1920.

—— Joseph Charles Arthur, in Lloyd, C. G., Mycological Notes (No. 62). 904–906. Je 1920.

—— Lentibulariaceae, in Britton, N. L., and Millspaugh, C. F., The Bahama Flora. 393–395. 26 Je 1920.

—— Bibliography (in part) in Britton, N. L., and Millspaugh, C. F., The Bahama Flora. 656–662. 26 Je 1920.

—— *Jeffersonia diphylla.* Addisonia **5**: 31, 32. *pl. 176.* Issue for 30 Je 1920.

—— (Biographical Notes), in Small, J. K., In quest of lost cacti. Jour. N. Y. Bot. Gard. **21**: 161–176. Issue for S 1920.

—— Lentibulariaceae, in Britton, N. L., Descriptions of Cuban plants new to science. Mem. Torrey Club **16**: 110, 111. 13 S 1920.

—— (Biographies of American medical botanists), in Kelly, H. A. and Burrage, W. L., American medical Biographies. 20, 21, 141, 142, 206, 207, 436, 558, 559, 649, 650, 669, 670, 1019, 1177, 1178. 1920.

—— Report of the Bibliographer (for 1919). Bull. N. Y. Bot. Gard. **10**: 265, 266. 30 Je' 1920.

Boynton, K. R. *Diplotaxis tenuifolia.* Addisonia **5**: 3, 4. *pl. 162.* Issue for 31 Mr 1920.

—— *Platycodon grandiflorum.* Addisonia **5**: 13, 14. *pl. 167.* Issue for 31 Mr 1920.

—— Vocational education in gardening for disabled and convalescent soldiers and sailors. Jour. N. Y. Bot. Gard. **21**: 87–94. Issue for My 1920.

———— *Monarda media.* Addisonia **5**: 39. *pl. 180.* Issue for 30 S 1920.

———— *Ceratostigma plumbaginoides.* Addisonia **5**: 45, 46. *pl. 183.* Issue for 30 S 1920.

———— Plants for Bird Tangle. The Florist's Exchange **50**: 349. 1920.

———— Report of the Supervisor of Gardening Instruction (for 1919). Bull. N. Y. Bot. Gard. **10**: 246–253. 30 Je 1920.

Britton, E. G. Mosses of Bermuda. Bryologist **22**: 87. 17 F. 1920.

———— Musci, in Britton, N. L., and Millspaugh, C. F., The Bahama Flora. 477–500. 26 Je 1920.

———— *Adlumia fungosa.* Addisonia **5**: 21, 22. *pl. 171.* Issue for 30 Je 1920.

———— The trailing arbutus. The Guide to Nature **13**: 62. S 1920.

———— Disappearing wild flowers. Torreya **20**: 101. Issue for S & O 1920.

———— Report of the Honorary Curator of Mosses (for 1919). Bull. N. Y. Bot. Gard. **10**: 271, 272. 30 Je 1920.

Britton, N. L. Flora of the District of Columbia. Torreya **19**: 244–246. 26 Ja 1920 (Review.)

———— The wild Pimento of Jamaica. Jour. N. Y. Bot. Gard. **21**: 38, 39. Issue for F 1920.

———— About Paulownia trees. Jour. N. Y. Bot. Gard. **21**: 72, 73. Issue for Ap 1920.

———— *Cephalanthus occidentalis.* Addisonia **5**: 17, 18. *pl. 171.* Issue for 30 Je 1920.

———— A botanical expedition to Trinidad. Jour N. Y. Bot. Gard. **21**: 101–118. Issue for Je 1920.

———— Descriptions of Cuban plants new to science (with assistance of P. Wilson.) Mem. Torrey Club. **16**: 57–118. 13 S 1920.

———— Two new West Indian plants. Torreya **20**: 83, 84. 20 S 1920.

———— George W. Perkins. Jour. N. Y. Bot. Gard. **21**: 179. Issue for S 1920.

———— Report of the Secretary and Director-in-Chief for the year 1919. Bull. N. Y. Bot. Gard. **10**: 213–229. 30 Je 1920.

Britton, N. L., and Millspaugh, C. F. The Bahama Flora. i–viii + 1695. New York. 26 Je 1920.

Britton, N. L., and Rose, J. N. The Cactaceae. Vol. II. i–vii + 1–237. *pl. 1–40* + *f. 1–304.* Washington, 1920.

Gleason, H. A. Organization of the American Iris Society. Jour. N. Y. Bot. Gard. **21**: 39, 40. Issue for F 1920.

———— Some applications of the quadrat method. Bull. Torrey Club **47**: 21–23. 18 F 1920.

———— *Vernonia crinita.* Addisonia **5**: 11, 12. *pl. 166.* Issue for 31 Mr 1920.

———— *Dracocephalum speciosum.* Addisonia **5**: 27, 28. *pl. 174.* Issue for 30 Je 1920.

———— The new greenhouse of the New York Botanical Garden. Museum Work **2**: 195. *pl. 1.* 1920.

———— The measurement of preglacial time. Science II. **52**: 340. 1920.

———— A new biological journal. Science II. **52**: 387, 388. 1920.

———— Report of the First Assistant (for 1919). Bull N. Y. Bot. Gard. **10**: 230–234. 30 Je 1920.

Harlow, S. H. Report of the Librarian (for 1919). Bull. N. Y. Bot. Gard. **10**: 267–269. 30 Je 1920.

Hollick, A. A graphical representation of geologic time. Sci. Am. **122**: 27. *f.* (*p. 31*). 10 Ja 1920.

———— Bird of Paradise plumage. Proc. Staten Is. Assoc. Arts and Sci. **7**: 1–4. 21 Je 1920.

———— Catskill aqueduct celebration exhibit. Ibid.: 5–12. 21 Je 1920.

———— *Quercus heterophylla* in the Clove Valley. Ibid.: 32–34. *pl. 4.* 21 Je 1920.

———— Wm. T. Davis' *Juniperus communis* on Long Island and Staten Island. Ibid.: 38. 21 Je 1920 (Review).

———— (Notes chiefly relating to Staten Island) Ibid.: 45, 46; 48, 49; 53, 54; 54–56; 56, 57; 58. 21 Je 1920.

———— Cape Cod has changed since the Pilgrims landed. N. Y. Evening Post (Magazine Section). 25 S 1920.

———— Report of the Honorary Curator of fossil plants (for 1919). Bull. N. Y. Bot. Gard. **10**: 272, 273. 30 Je 1920.

Howe, M. A. Observations on monosporangial discs in the genus *Liagora.* Bull. Torrey Club **47**: 1–8. *pl. 1* + *f. 25–29.* 18 F 1920.

———— Ricciaceae, in Britton, N. L., and Millspaugh, C. F., The Bahama Flora. 502, 503. 26 Je 1920.

———— Algae, in Britton, N. L., and Millspaugh, C. F., The Bahama Flora. 553–618. 26 Je 1920.

———— The 1920 dahlia border. Jour. N. Y. Bot. Gard. 21: 138. Issue for Je 1920.

Murrill, W. A. Corrections and additions to the polypores of temperate North America. Mycologia 12: 6–24. 2 F 1920.

———— Fungi from Hedgcock. Mycologia 12: 41, 42. 2 F 1920.

———— Collecting fungi at Yama Farms. Mycologia 12: 42, 43. 2 F 1920.

———— *Trametes serpens.* Mycologia 12: 46, 47. 2 F 1920.

———— The genus *Poria.* Mycologia 12: 47–51. 2 F 1920.

———— Collecting fungi near Washington. Mycologia 12: 51, 52. 2 F 1920.

———— Saturday and Sunday spring lectures, 1920. Jour. N. Y. Bot. Gard. 21: 54, 55. Issue for Mr 1920.

———— Illustrations of fungi—XXXII. Mycologia 12: 59–61. *pl. 2.* 8 Ap 1920.

———— Light-colored resupinate polypores—I. Mycologia 12: 77–92. 8 Ap 1920.

———— *Polyporus excurrens* Berk. & Curt. Mycologia 12: 107, 108. 8 Ap 1920.

———— A correction (on *Poria*). Mycologia 12: 108, 109. 8 Ap 1920.

———— *Daedalea extensa* rediscovered. Mycologia 12: 110, 111. 8 Ap 1920.

———— The artist's bracket fungus. Sci. Am. 122: 365. 1920. (Illust.)

———— What is cane sugar? Sci. Am. 122: 399. 10 Ap 1920. (Illust.)

———— Plant growths that shed light. Sci. Am. 122: 427, 440. 17 Ap 1920. (Illust.)

———— The dendrograph—an instrument that keeps tabs on tree growth. Sci. Am. 122: 595. 29 My 1920. (Illust.)

———— Trees girdled by meadow mice. Jour. N. Y. Bot. Gard. 21: 94–97. Issue for My 1920.

———— Where chocolate comes from (and its manufacture). Sci. Am. 122: 626. 5 Je 1920. (Illust.)

———— Another new truffle. Mycologia **12:** 157, 158. *f. 1.* 5 Je 1920.

———— Pier Andrea Saccardo. Mycologia **12:** 164. 5 Je 1920.

———— A mycologist in the making. Mycologia **12:** 165. 5 Je 1920.

———— Kauffman's Agaricaceae. Mycologia **12:** 166. 5 Je 1920.

———— Oudeman's work on fungi. Mycologia **12:** 169. 5 Je 1920.

———— Fighting for healthy plants. Sci. Am. **122:** 677. 19 Je 1920.

———— Autobasidiomycetes, in Britton, N. L., and Mills-paugh, C. F., The Bahama Flora. 637–645. 26 Je 1920.

———— A new *Amanita.* Mycologia **12:** 291, 292. 4 S 1920.

———— Many odd plants kin to pineapple. The Sun and New York Herald. 12 S 1920. (Illust.)

———— Mushrooms, good and evil (13 articles). Evening Post, Sep., Oct. 1920. (Illust.)

———— The blue beetle of the willow. Country Life **38:** 114–132. O 1920. (Illust.)

———— How meadow mice destroy trees. Sci. Am. **122:** 525–536. 20 N 1920. (Illust.)

———— Plants as an inspiration in the art of early peoples. Garden Mag. **32:** 246–248. D 1920. (Illust.)

———— Light-colored resupinate polypores—II. Mycologia **12:** 299–308. 27 D 1920.

———— The fungi of Blacksburg, Virginia. Mycologia **12:** 322–328. 27 D 1920.

———— Notes (on literature and events). Mycologia **12:** 36–40; 104–107; 159–164; 286–289; 334–342. 1920.

———— Report of the Supervisor of Public Instruction (for 1919). Bull. N. Y. Bot. Gard. **10:** 240–246. 30 Je 1920.

Nash, G. V. Hardy woody plants in the New York Botanical Garden. Jour. N. Y. Bot. Gard. **21:** 56–60; 74–77; 119–124. 1920.

———— *Viburnum dilatum.* Addisonia **5:** 1, 2. *pl. 161.* Issue for 31 Mr 1920.

———— Rosa "Dr. Van Fleet." Addisonia **5:** 7, 8. *pl. 164.* Issue for 31 Mr 1920.

———— *Amygdalus Davidiana.* Addisonia **5:** 9, 10. *pl. 165.* Issue for 31 Mr 1920.

———— *Benzoin aestivale.* Addisonia **5:** 15, 16. *pl. 168.* Issue for 31 Mr 1920.

———— *Corylopsis spicata.* Addisonia **5:** 19, 20. *pl. 170.* Issue for 30 Je 1920.

———— *Aphelandra nitens.* Addisonia **5:** 23, 24. *pl. 172.* Issue for 30 Je 1920.

———— *Crataegus Phaenopyrum.* Addisonia **5:** 33. *pl. 177.* Issue for 30 S 1920.

———— *Viburnum Sieboldii.* Addisonia **5:** 35, 36. *pl. 178.* Issue for 30 S 1920.

———— *Stephanandra Tanakae.* Addisonia **5:** 37. *pl. 179.* Issue for 30 S 1920.

———— *Clethra barbinervis.* Addisonia **5:** 41. *pl. 181.* Issue for 30 S 1920.

———— *Solidago rugosa.* Addisonia **5:** 43–44. *pl. 182.* Issue for 30 S 1920.

———— Report of the Head Gardener (for 1919). Bull. N. Y. Bot. Gard. **10:** 253–259. 30 Je 1920.

Pennell, F. W. Scrophulariaceae of the local Flora.—V. Torreya **19:** 235–242. 26 Ja 1920.

———— Scrophulariaceae of the southeastern United States. Proc. Acad. Nat. Sci. Phila. **71:** 224–291. 11 Mr 1920.

———— Soil preferences of Scrophulariaceae. Torreya **20:** 10, 11. 1 Ap 1920.

———— Scrophulariaceae of the central Rocky Mountain states. Cont. U. S. Nat. Herb. **20:** 313–381. 30 Ap. 1920.

———— Scrophulariaceae, in Britton, N. L., Descriptions of Cuban plants new to science. Mem. Torrey Club **16:** 103–106. 13 S 1920.

———— Scrophulariaceae of Colombia—I. Proc. Acad. Nat. Sci. Phila. **72:** 136–188. 4 O 1920.

Rusby, H. H. Suggestions for the revision of the U. S. Pharmacopoeia. Jour. Am. Pharm. Assoc. **9:** 240–249. 1920.

———— Codes of nomenclature in the U. S. Pharmacopoeia. Jour. Am. Pharm. Assoc. **9:** 670, 671. 1920.

———— The Mulford Biological Expedition of the Amazon Basin. Oil, Paint and Drug Reporter **98:** 27. 18 O 1920.

———— Descriptions of three hundred new species of South American plants with an index to all previously described species from South America by the same author. New York. 1–170. D 1920.

———— Report of the Honorary Curator of the economic collections (for 1919). Bull. N. Y. Bot. Gard. 10: 269–271. 30 Je 1920.

Rusk, H. M. The effect of zinc sulphate on protoplasmic streaming. Bull. Torrey Club 47: 425–431. *f. 1, 2.* 20 O 1920.

Rydberg, P. A. (Rosales) Fabaceae: Psoraleae. N. A. Fl. 24: 65–136. 27 Ja 1920.

———— Notes on Rosaceae—XII. Bull. Torrey Club **47:** 45–66. 10 Mr 1920.

———— *Bembicidium* gen. nov., in Britton, N. L., Descriptions of Cuban plants new to science. Mem. Torrey Club. **16:** 68–69. 13 S 1920.

———— Phytogeographical notes on the Rocky Mountain region—IX. Bull. Torrey Club **47:** 441–454. 29 O 1920.

———— Henry and Flood's The Douglas Fir. Torreya **20:** 102–104. 12 N 1920

Seaver, F. J. Photographs and descriptions of cup-fungi— VIII. Mycologia **12:** 1–5. *pl. 1.* 2 F 1920.

———— Notes on North American Hypocreales—IV. Mycologia **12:** 93–98. *pl. 6.* 8 Ap 1920.

———— Fungi, in Britton, N. L. and Millspaugh, C. F., The Bahama Flora. 631–645. 26 Je 1920.

Small, J. K. *Pieris floribunda.* Addisonia 5: 5, 6. *pl. 163.* Issue for 31 Mr 1920.

———— Of grottoes and ancient dunes. Jour. N. Y. Bot. Gard. **21:** 25–38. *pl. 241, 242.* Issue for F 1920; 45–54. *pl. 243, 244.* Issue for Mr 1920.

———— Cypress and population in Florida. Jour. N. Y. Bot. Gard. **21:** 81–86. *pl. 245–247.* Issue for May 1920.

———— The land of ferns. Jour. Elisha Mitchell Sci. Soc. **35:** 92–104. *pl. 24–28.* Je 1920.

———— *Hydrangea quercifolia.* Addisonia 5: 29, 30. *pl. 175.* Issue for Je 1920.

———— In quest of lost cacti. Jour. N. Y. Bot. Gard. **21:** 161–178. *pl. 251, 252.* Issue for S 1920.

———— *Grossularia curvata.* Addisonia **5**: 47–48. *pl. 184.* Issue for 30 S 1920.

———— Report of the Head Curator of the Museums and Herbarium (for 1919). Bull. N. Y. Bot. Gard. **10**: 234–239. 30 Je 1920.

Stout, A. B. The aims and methods of plant breeding. Jour. N. Y. Bot. Gard. **21**: 1–16. Issue for Ja 1920.

———— Further experimental studies of self-incompatibility in hermaphrodite plants. Jour. Genet. **9**: 5–129. *pl. 3, 4.* Ja 1920.

———— Conference notes for December (1919). Jour. N. Y. Bot. Gard. **21**: 16–18. Issue for Ja 1920;—January (1920). 41, 42. Issue for F 1920;—February. 60–62. Issue for Mr 1920;—March. 78. Issue for Ap 1920;—April. 97, 98. Issue for My 1920.

———— Chinese Pe-tsai, the salad Mogul. Garden Mag. **31**: 368–369. Au 1920. (Illust.)

———— Report of the Director of the Laboratories (for 1919). Bull. N. Y. Bot. Gard. **10**: 259–262. 30 Je 1920.

Watson, E. E. *Corylus rostrata.* Addisonia **5**: 25, 26. *pl. 175.* Issue for 30 Je 1920.

Williams, R. S. *Grimmia brevirostris.* Bryologist **23**: 52. *pl. 3.* 17 S 1920.

———— Calymperaceae of North America. Bull. Torrey Club **47**: 367–396. *pl. 15–17.* 20 O 1920.

———— *Sematophyllum Smallii.* Bryologist **23**: 76–78. *pl. 6.* D 1920.

Wilson, P. [Assisted at many points] in Britton, N. L., Descriptions of Cuban plants new to science. Mem. Torrey Club **16**: 57–118. 13 S 1920.

SPRING AND SUMMER LECTURES

An attractive series of public lectures on botanical and horticultural subjects will be given in the Lecture Hall of the Museum Building on Saturday and Sunday afternoons at four o'clock as listed below. Most of them will be illustrated by lantern slides. In addition to the two floral exhibits here announced, there will be one early in June devoted to roses, the exact date of which can not be fixed at this time.

April 30. Botanizing on a Volcano, Dr. H. A. Gleason.

May 1. Spring Flowers, Dr. N. L. Britton.

May 7. The Rôle of Plants in Man's Evolution, Dr. W. A. Murrill.

May 8. Window Gardens for City Homes. Mr. Van Evrie Kilpatrick.

May 14. Mountaineering in the Northwest, Mr. Leroy Jeffers. (Exhibition of flowers, May 14 and 15)

May 15. Plant Hybrids, Dr. A. B. Stout.

May 21. Water-lilies and How to Grow Them, Mr. G. V. Nash.

May 22. Practical Hints for Home Gardeners, Mr. Hugh Findlay.

May 28. Bog Vegetation at Cranberry Lake, Prof. W. L. Bray.

May 29. Why Study Nature?, Dr. W. A. Murrill.

June 4. Sea Gardens of the Tropics, Dr. M. A. Howe.

June 5. Collecting in the Rocky Mountains, Dr. F. W. Pennell.

June 11. How Nature Scatters Seeds, Dr. G. C. Fisher.

June 12. The Origin of Plant Life, Dr. W. A. Murrill.

June 18. Destructive Insects, Dr. F. J. Seaver.

June 19. The Vegetation of Long Island, Mr. Norman Taylor.

June 25. Books on Gardening, Dr. J. H. Barnhart.

June 26. Color in the Garden, Miss Emily Exley.

July 2. Floral and Scenic Features of Haïti, Mr. G. V. Nash.

July 3. The Glory of the Annuals, Mr. Arthur Herrington.

July 9. The Golden Mean in Nature, Dr. W. A. Murrill

July 10. Planning and Planting the Home Grounds, Mr. H. D. Hemenway.

July 16. Immunity to Disease in Plants, Dr. W. A. Murrill.

July 17. Wild Flowers, Dr. H. D. House.

July 23. Some American Botanists, Dr. J. H. Barnhart.

July 24. The Significance of Sex in Plants, Dr. W. A. Murrill.

July 30. Summer Garden Flowers, Mr. G. V. Nash.

July 31. How Plant Life Will Cease, Dr. W. A. Murrill.

August 6. Botanical Cruises in the Bahamas, Dr. M. A. Howe.

August 7. The Care of the Vegetable Garden, Mr. Hugh Findlay.

August 13. Impressions of the Colombian Andes, Dr. F. W. Pennell.

August 14. Some of Our Common Food Plants. Dr. M. T. Cook.

August 20. Evergreens and How to Grow Them, Mr. G. V. Nash.

(Exhibition of Gladioli, August 20 and 21)

August 21. The Position of Plants in the System of Nature, Dr. W. A. Murrill.

August 27. Colorado, Dr. F. J. Seaver.

August 28. The Origin of Cultivated Plants, Dr. A. B. Stout.

The Museum Building is reached by the Harlem Division of the New York Central and Hudson River Railroad to Botanical Garden Station, by trolley cars to Bedford Park, or by Third Avenue Elevated Railway to Botanical Garden, Bronx Park. Visitors coming by the Subway change to the Elevated Railway at 149th Street and Third Avenue. Those coming by the New York, Westchester and Boston Railway change at 180th Street for crosstown trolley, transferring north at Third Avenue.

W. A. MURRILL

CONFERENCE NOTES FOR MARCH

The March conference of the Scientific Staff and Registered Students of the Garden was held in the museum building on Wednesday, March 2, 1921, at 3:30 P. M.

The program was as follows: "Remarks on Fossil Algae" by Dr. Marshall A. Howe and "Rubber Content of North American Plants" by Dr. H. M. Hall.

Dr. Howe reviewed the few papers in which American fossil algae have been described, and exhibited specimens of "nullipore limestone" from Trinidad and Santo Domingo, sent for determination by Dr. T. Wayland Vaughan of the U. S. Geological Survey. Two specimens from Trinidad (Lower Miocene), of which photomicrographs were exhibited, were considered to represent two undescribed species.

A fossil organism of a different character was recently collected by Mr. William J. Sinclair of Princeton University, in the fresh-water Oligocene of the "Bad Lands" of South Dakota. It forms extended, laminated, calcareous crusts or smaller concretions with concentric lamellations. Its microscopic cellular structure is very imperfectly preserved and a student of modern algae would hardly feel justified in going further than to say that this South Dakota fossil probably represents the remains of some lime-depositing blue-green alga comparable with some now living in certain lakes of central New York and in Little Conestoga Creek, Lancaster County, Pa.

The speaker discussed especially fossil organisms from the Cambrian and Pre-cambrian that have been referred to the algae by Dr. C. D. Walcott and also other ancient fossils of a probably algal nature that have been known under the generic name *Cryptozoon* as well as others of more doubtful affinities that constitute the genus *Solenopora*.

Dr. H. M. Hall spoke very interestingly on the rubber content of various native North American plants, a matter which Dr. Hall with associates, has had under special investigation during the past few years. It has been found that rubber of good quality can be obtained from various species of *Chrysothamnus* and *Haplopappus* of the Aster Family of plants. One species, *Chrysothamnus nauseosus*, commonly called rabbit-bush, is a large shrub that grows abundantly over the semi-desert areas of Colorado, Nevada and Utah. The present stand of the varieties of this species would yield, it is estimated, about 300,000,000 lbs. of good rubber. This rubber could not be obtained at a price that could compete with the rubber produced from the tree, *Hevea braziliensis*, which is grown in the tropics, but under the stress of national emergency in war time conditions, the supply available might become a valuable asset.

The possibility of improving native rubber-bearing plants is being considered. The rabbit-bush can be grown without irrigation on waste and alkaline lands of the western states. Some varieties of it withstand winter temperatures of −20° F., and other varieties endure extreme summer heat. The plant appears to be a most promising species for use in any attempts that are to be made in the establishment of a rubber-growing industry in the United States.

Dr. Hall also reported on preliminary studies of the rubber content in many other species, especially of the Dogbane and Milkweed families.

A. B. STOUT
Secretary of the Conference

NOTES, NEWS AND COMMENT

A guide to nature-study and school garden opportunities in New York City, prepared by a committee representing the local institutions and agencies interested in nature and garden education and edited by Van Evrie Kilpatrick, was recently published by the School Garden Association of New York. This interesting and valuable publication informs superintendents, principals, and teachers regarding the resources for nature-education in our city, which are greater than those of any other city in the world. It explains where and how teachers may get instruction, materials, charts, pictures, lantern slides, equipment, and literature for their daily teaching; shows where teachers may take courses in science, nature, and gardening; where they may attend special lectures on these subjects; and how such lectures may be secured for a given school or community. It describes how pupils may work effectively, how classes may visit institutions, parks, greenhouses, gardens, fields, and many other points of nature contact; how they may use school grounds, make nature collections, use the public libraries, work in school gardens, get nature material, see nature exhibits, and secure credits through awards. The cordial spirit of cooperation that has characterized this undertaking from the beginning is a source of great promise for the future growth of true nature education. (W. A. M.)

An underground gasteromycete, apparently a species of *Hysterangium*, was brought in about the middle of February by Mr. L. Robba, who collected it with a trained truffle dog under an oak tree near White Plains, New York. The soil was not frozen hard because of the mild weather and a layer of two or three inches of leaves, but the tiny "puffballs" were frozen and made very poor specimens when dried. The spores were rather rough, ovoid, and distinctly umber-brown under a microscope. Mr. Robba naturally thought they were truffles, but he did not notice any odor, and he recalled that his dog was not particularly "interested" in the find, only scratching a little to mark the spot and then walking away. The plants were unearthed by scraping off the covering of leaves and digging about two inches into the soil. There must have been *some* odor present, otherwise the dog would not have been attracted. (W. A. M.)

In the central display house, conservatory range 2, there was a fine display of flowering bulbs and other plants for Easter. Daffodils, tulips, hyacinths, hippeastrums, and squills lent a great show of color. Other flowers were *Primula malacoides* and its white variety, *P. obconica*, and *P. kewensis*, azaleas and Forsythias. This display has been continued for some weeks.

The large bottle-brush tree, *Callistemon citrinus*, is in full bloom in the central display house, conservatory range 2. It is covered with hundreds of its cylindric, bright red flower-clusters, from which the tree derives its name. There are other plants of *Callistemon* also coming into bloom. *Acacia pulchella* and *A. hispidissima* are covered with a wealth of bright yellow blossoms, the last of the species in the collection to flower.

Out-of-doors everything is one to two weeks in advance of the usual time, owing to the wonderful spring-like weather which has prevailed, some days the heat even touching that of summer. It was necessary to uncover the large collections of tulips about two weeks ahead of the usual time to prevent the plants drawing too much. The Forsythias form masses of gold all over the grounds, great masses greeting the eye in all directions. They are unusually fine this season, the group of *Forsythia intermedia* near the elevated railroad approach and the large mass near the Harlem Railroad depot being especially imposing. In the flower beds at conservatory range 1 the Siberian squills seems never to have been so blue before, and the vividness of the glory-of-the-snow, *Chionodoxa Luciliae*, is striking. Crocuses, purple, lilac, yellow and white, are in abundance, and the daffodils are just beginning to show their blossoms.

In the fruticetum are more masses of *Forsythia*, rivalling those to which allusion has already been made. And nearby are the Cornelian cherry, *Cornus Mas*, and the Sandzaki, *Cornus officinalis*, a Japanese relative of the other, the latter, if anything, the more striking of the two. That sweetest of the early honeysuckles, *Lonicera fragrantissima*, is in full bloom, not only here but in many other parts of the grounds. *Magnolia Kobus* is a mass of bloom, and the willows nearby are sending out myriads of "pussies." (March 29, G. V. N.)

It may be of interest to know that the following species were in bloom in the Botanical Garden grounds on or before March 21st.

Crocuses
Siberian Squill
Glory-of-the-snow
Honeysuckles, two species
Dogwood, two species
David's Peach
Buffalo-berry

Forsythia, two species
Spice-bush
Coltsfoot
Mountain Spurge
Adonis amurensis
Jasminum nudiflorum

Miss Wakefield, the well-known Mycologist of Kew Gardens, England, arrived in New York on March 10, after spending the winter collecting in the British West Indies. On March 19, she left for a tour through parts of the Eastern United States, and will sail for England in May. Her chief interest at the Garden was the large collection of polypores from the American tropics.

ACCESSIONS

LIBRARY, FROM NOV. 1, 1920, to JAN. 31, 1921

ARBER, AGNES. *Water plants: a study of aquatic angiosperms.* Cambridge, 1920. (Given by Mrs. Theron G. Strong.)

BERKENHOUT, JOHN. *Synopsis of the natural history of Great-Britain and Ireland.* 2 vols. London, 1789.

BRADLEY, RICHARD. *History of succulent plants.* London, 1716–1727.

BRUCH, PHILIPP, SCHIMPER, WILHELM PHILIPP, & GUMBEL, WILHELM THEODOR. *Bryologia europaea.* 6 vols. Stuttgartiae. 1836–55.

CURTIS, WILLIAM. *Companion to the Botanical magazine.* London, 1788.

CURTIS, WILLIAM. *Subscription catalogue of the Brompton botanic garden for . . . 1792.* London, 1792.

CZAPEK, FRIEDRICH. *Biochemie der Pflanzen.* Ed. 2. vol. 2. Jena, 1920.

DRYANDER, JONES. *Catalogue bibliothecae historico-naturalis Josephi Banks.* 5 vols. Londini, 1798–1800.

FIELD, BARRON. *Geographical memoirs on New South Wales by various hands.* London, 1825.

FLINDERS, MATTHEW. *Voyage to Terra Australis . . . in . . . 1801, 1802 & 1803 in his majesty's ship The Investigator.* 3 vols. London, 1814.

FORBES, JAMES. *Pinetum woburnense.* London, 1839.

GIBSON, GEORGE STACEY. *Flora of Essex.* London, 1862.

GORDON, GEORGE. *The pinetum.* Ed. 3. London, 1880.

GRIGOR, JAMES. *The eastern arboretum.* London, 1841.

GRINDON, LEOPOLD HARTLEY. *British and garden botany.* London, 1864.

HEDRICK, ULYSSES PRENTISS. ed. *Sturtevant's notes on edible plants.* Albany, 1919. (By exchange with New York Agricultural Experiment Station.)

HILL, JOHN. *Exotic botany illustrated, in thirty-five figures of curious and elegant plants.* London, 1759.

HOOKER, WILLIAM JACKSON. *Description of Victoria regia.* London, 1847.

HOUSE, HOMER DOLIVER. *Wild flowers of New York.* 2 vols. Albany, 1918 (Given by Dr. N. L. Britton.)

International horticultural exhibition and botanical congress held in London from May 22–31, 1866. London, 1867.

KNIGHT, THOMAS ANDREW. *A selection from the physiological and horticultural papers . . . in the Transactions of the Royal and Horticultural societies.* London, 1841.

LABILLARDIERE, JACQUES JULIEN HOUTON DE. *Relation du voyage à la recherche de la Pérouse, pendant les années 1791, 1792.* 2 vols. Paris. An 8.

LAWSON, PETER. *Pinetum britannicum.* 3 vols. Edinburgh. 1884.

LEROY, ANDRE. *Dictionnaire de pomologie.* 6 vols. in 4. Paris, 1867–77.

LINDLEY, JOHN. *Pomologia britannica.* 3 vols. London, 1841.

LOUDON, JANE WELLS. *The ladies' flower-garden of ornamental perennials.* 2 vols. London, 1843, 44.

LOUDON, JOHN CLAUDIUS. *Encyclopaedia of trees and shrubs.* 2 vols. London, 1842.

MANGLES, JAMES. *Floral calendar, monthly and daily.* London, 1839.

OUDEMANS, CORNEILLE ANTOINE JEAN ABRAM. *Enumeratio systematica fungorum.* Vols. 1 & 2. Hagae Comitum, 1919–20.

PEREIRA, JONATHAN. *Elements of materia medica and therapeutics.* Ed. 2. 2 vols. London, 1842.

PETERS, WILHELM, CARL HARTWIG. *Naturwissenschaftliche Reise nach Mossambique . . . in 1842 bis 1848 ausgeführt. Botanik.* 2 vols. Berlin, 1862–64.

PULTENEY, RICHARD. *A general view of the writings of Linnaeus.* Ed. 2. London, 1805.

REDOUBTE, PIERRE JOSEPH. *Choix des plus belles fleurs prises dans différentes familles du règne végétal.* Paris, 1827.

RONALDS, HUGH. *Pyrus malus brentfordiensis.* London, 1831.

ROSCOE, MRS. EDWARD. *Floral illustrations of the seasons.* London, 1831.

SAVI, GAETANO. *Flora italiana . . . raccolta delle piante . . . che si coltivano nei giardini d'Italia.* 3 vols. Pisa, 1818–34.

SCHIMPER, WILHELM PHILIPP. *Musci europaei novi; vel, Bryologiae europaeae supplementum.* Stuttgartiae, 1864–66.

SCHIMPER, WILHELM PHILIPP. *Corallarium Bryologiae europaeae.* Stuttgartiae, 1855.

SMEE, ALFRED. *The potatoe plant.* London, 1846.

SOWERBY, JOHN EDWARD. *British wild flowers . . . described by C. Pierpont Johnson.* Lambeth, 1860.

THOMSON, ANTHONY TODD. *Lectures on the elements of botany.* part 1. London, 1822.

TUSSAC, F. RICHARD DE. *Flora antillarum.* 4 vols. Parisiis, 1808–27.

VERSCHAFFELT, AMBROISE COLETTE ALEXANDRE. *Nouvelle inconographie des Camellias.* 13 vols. in 6. Gand, 1848–60.

PUBLICATIONS OF

The New York Botanical Garden

Journal of the New York Botanical Garden, monthly, illustrated, containing notes, news, and non-technical articles of general interest. Free to all members of the Garden. To others, 10 cents a copy; $1.00 a year. [Not offered in exchange.] Now in its twenty-second volume.

Mycologia, bimonthly, illustrated in color and otherwise; devoted to fungi, including lichens; containing technical articles and news and notes of general interest, and an index to current American mycological literature. $4.00 a year; single copies not for sale. [Not offered in exchange.] Now in its thirteenth volume.

Addisonia, quarterly, devoted exclusively to colored plates accompanied by popular descriptions of flowering plants; eight plates in each number, thirty-two in each volume. Subscription price, $10.00 a year. [Not offered in exchange.] Now in its sixth volume.

Bulletin of the New York Botanical Garden, containing the annual reports of the Director-in-Chief and other official documents, and technical articles embodying results of investigations carried out in the Garden. Free to all members of the Garden; to others, $3.00 per volume. Now in its tenth volume.

North American Flora. Descriptions of the wild plants of North America, including Greenland, the West Indies, and Central America. Planned to be completed in 34 volumes. Roy. 8vo. Each volume to consist of four or more parts. Subscription price, $1.50 per part; a limited number of separate parts will be sold for $2.00 each. [Not offered in exchange.]

Vol. 3, part 1, 1910. Nectriaceae—Fimetariaceae.

Vol. 7, part 1, 1906; part 2, 1907; part 3, 1912; parts 4 and 5, 1920; part 6, 1921. Ustilaginaceae—Aecidiaceae (pars). (Parts 1 and 2 no longer sold separately.)

Vol. 9 (now complete), parts 1-7, 1907-1916. Polyporaceae—Agaricaceae (pars). (Parts 1-3 no longer sold separately.)

Vol. 10, part 1, 1914; parts 2 and 3, 1917. Agaricaceae (pars).

Vol. 15, parts 1 and 2, 1913. Sphagnaceae—Leucobryaceae.

Vol. 16, part 1, 1909. Ophioglossaceae—Cyatheaceae (pars).

Vol. 17, part 1, 1909; part 2, 1912; part 3, 1915. Typhaceae—Poaceae (pars).

Vol. 21, part 1, 1916; part 2, 1917; part 3, 1918. Chenopodiaceae—Allioniaceae.

Vol. 22, parts 1 and 2, 1905; parts 3 and 4, 1908; part 5, 1913; part 6, 1918. Podostemonaceae—Rosaceae.

Vol. 24, part 1, 1919; part 2, 1920. Fabaceae (pars).

Vol. 25, part 1, 1907; part 2, 1910; part 3, 1911. Geraniaceae—Burseraceae.

Vol. 29, part 1, 1914. Clethraceae—Ericaceae.

Vol. 32, part 1, 1918. Rubiaceae (pars).

Vol. 34, part 1, 1914; part 2, 1915; part 3, 1916. Carduaceae—Anthemideae.

Memoirs of the New York Botanical Garden. Price to members of the Garden, $1.50 per volume. To others, $3.00. [Not offered in exchange.]

Vol. I. An Annotated Catalogue of the Flora of Montana and the Yellowstone Park, by Per Axel Rydberg. ix + 492 pp., with detailed map. 1900.

Vol. II. The Influence of Light and Darkness upon Growth and Development, by D. T. MacDougal. xvi + 320 pp., with 176 figures. 1903.

Vol. III. Studies of Cretaceous Coniferous Remains from Kreischerville, New York, by A. Hollick and E. C. Jeffrey. viii + 138 pp., with 29 plates. 1909.

Vol. IV. Effects of the Rays of Radium on Plants, by Charles Stuart Gager. viii + 278 pp., with 73 figures and 14 plates. 1908.

Vol. V. Flora of the Vicinity of New York: A Contribution to Plant Geography, by Norman Taylor. vi + 683 pp., with 9 plates. 1915.

Vol. VI. Papers presented at the Celebration of the Twentieth Anniversary of the New York Botanical Garden. viii + 592 pp., with 43 plates and many text figures. 1916.

Contributions from the New York Botanical Garden. A series of technical papers written by students or members of the staff, and reprinted from journals other than the above. Price, 25 cents each. $5.00 per volume. In the ninth volume.

NEW YORK BOTANICAL GARDEN
Bronx Park, New York City

Vol. XXII   July, 1921   No. 259

JOURNAL

OF

The New York Botanical Garden

EDITOR

R. S. WILLIAMS

Administrative Assistant

CONTENTS

PRICE $1.00 A YEAR; 10 CENTS A COPY

PUBLISHED FOR THE GARDEN
AT 8 WEST KING STREET, LANCASTER, PA.
INTELLIGENCER PRINTING COMPANY

PLATE 258

Zamia integrifolia in pine woods on the Everglade Keys. Plant with ovulate (large) cones about the middle, one with staminate (smaller) cone near the right side of plate. Many acres of the rocky pineland were formerly as abundantly clothed with this cycad as is shown in plate. Large areas are as yet copiously populated with this esculent; still, in many places the orginal abundant growth has been much reduced by reckless harvesting, and in others, exterminated by the clearing of the land. The leaves are gathered, prepared, and used in floral decorations.

JOURNAL

OF

The New York Botanical Garden

| VOL. XXII | July, 1921 | No. 259 |

SEMINOLE BREAD — THE CONTI

A History of the Genus Zamia in Florida

WITH PLATES 258 AND 259

In the annals of Florida there is one wild esculent plant mentioned oftener than all the rest. This is the Conti of the Seminoles, the "Florida arrow-root" of the white man. Conti, in the Muskogee, of which the present-day Seminole speech of Florida is a varient, signifies "flour-root." Sometimes the adjective hateka or hatkee (white) is added to differentiate it from other starch-producing plants; but Conti is applied only to members of the genus *Zamia*.

Like many aboriginal or Indian names, the one in question appears in a variety of forms: Conti, Coontie, Coontia, Compte, Comtie, Koonti or Koontee. There are others. But in some such form the word dots the chronicles of Florida as references to "corn" run through the history of the world at large. And now this beautiful cycad that furnished a staple supply of starch to Florida's aborigines, then to her Seminoles, and later to the white population, is in danger of extermination.

There are a number of destructive agencies at work. The manufacture of "Florida arrowroot," the conversion of aboriginal shell-middens into material for new roads, the levelling of sand-dunes for building-lots, the clearing of land generally— these are but a few of them, that, if continued as at present, will soon relegate the once wide-spread zamia to a dwindling career in conservatories and gardens.

In this connection it is recorded that "from the tubers of this plant the Florida arrow-root is made. It is abundant in the southern part of the State. The tubers are large, frequently a foot long and three inches in diameter, rough and dark on the outside, but white inside and yield a large percentage of starch. It possesses an acid, poisonous ingredient which has to be washed out in the process of converting the root into starch. The Indians of the Everglades consume a great deal of starch as food, prepared by their rude processes, and also sell some, but it is inferior to that prepared by Americans with improved machinery."[1]

"In Florida this Cycad [*Zamia intergrifolia*] is largely cultivated for the sake of the starch contained in its roots, large quantities of the starch being made for the Key West and West Indian markets. There are several establishments now engaged in this industry upon the shores of Bay Bicayne and upon Miami River just below where it flows out from the Florida Everglades." . . [2]

"On the low grounds of Southern Florida grows a beautiful plant of the Cycadaceae, *Zamia integrifolia* Willd., the 'Coontie' of the Indians, which contains in the caudex and roots about 65 per cent of an excellent starch sometimes called Florida arrow-root. This plant supplied the Seminoles with food during their long wars with the United States and is now more or less cultivated."[3]

How early the cycads of Florida were referred to in the writings and records of the pioneer travelers we do not know, but the technical botanical history begins with the printing of the formal description of one of the species in 1789.

The exact origin, and the circumstances connected with the independent discoveries, of a rather odd, but common, plant of Florida seemed, until recently destined to remain mysteries, although its history begins but little over a century ago. Even the original collectors are not mentioned in connection with the type specimens or with the original descriptions of the species and until quite recently there seemed to be no record extant from which we might learn their identity.

For the past score of years two well-marked species of the genus *Zamia* have been known from Florida.

It is recorded that "reseaches have shown that there are at least two species of Zamia in Florida, where only one has heretofore been recognized as occurring. These are *Zamia floridana* DC. and *Z. pumila* L. It was found that neither of the forms

---

[1] Edward Palmer, American Naturalist 12: 600. 1878.

[2] Charles Sprague Sargent, Gardener's Chronicle. N. S. **26**: 146. 1886.

[3] Valery Havard, Bulletin of the Torrey Botanical Club **22**: 107. 1895.

studied could be referred to *Z. integrifolia* Ait. as has been done heretofore, this being a very distinct West Indian species."[1]

A note inspired by the paper in which the foregoing quotation occurs, follows: "The Cycads of Florida have recently been found . . . to include with certainty two species, *Zamia floridana* D C. and *Z. pumila* L. It is hence not known that Z. *angustifolia* Jacquemont and *Z. integrifolia* Aiton, are represented in Florida at all, or if indeed these latter are two distinct species. Nothing short of the comparison of plants from many localities, both insular and from the mainland well to the south, will settle this point, and also determine whether any other species than *Z. floridana* and *pumila* are indigenous to Florida."[2]

The wilderness of Florida served as the type locality for two species of *Zamia*—*Z. integrifolia* (1789) and *Z. floridana* (1868). The locality given for one species when first published was "East Florida," for that of the other merely "Florida."

Geography and imagination have played important and interesting parts in working out the facts and the history of *Zamia* in Florida. The lack of information above referred to and the localities "Florida" and "East Florida" were the inciting factors that led to the solution of the problems.

In the first place, Florida represents extensive territory, and as far as we know, no specimens of *Zamia* have been found growing wild in what has commonly been known as East Florida, for more than a century, unless the ones upon which the two species referred to above were founded. It is possible that the original specimens of the two species were secured from native plants in East Florida or from plants that were cultivated for their starchy stems by the Indians, all the supply of which might have been exterminated by extreme cold weather that is said to have ravished Florida in the thirties of the last century.

In the second place, "East Florida" might have been extended somewhat southward in the records of a collector to include the northern geographic limit of the genus *Zamia*, as we know it to-day, or it might have been used in the sense of peninsular Florida or of merely the eastern coast of Florida. In this latter case a coast line of some five hundred and fifty miles would have to be

[1] Herbert John Webber, Bulletin of the U. S. Department of Agriculture 1: 81. 1901.

[2] George Reber Wieland, American Journal of Science. Ser. IV. 13: 332. 1902.

considered. Thus, a very simple geographic statement may stimulate the imagination.

Although the finding of a record of where and by whom the type specimens of *Zamia integrifolia* and *Z. floridana* were collected seemed hopeless, we fortunately had access to the type-specimens themselves, though indirectly. We recently received tracings of both the specimens and the notes preserved on the original herbarium sheets. One of these, *Zamia integrifolia*, is at the British Museum in London; the other, *Z. floridana*, is in the De Candolle herbarium at Geneva.

Curiously enough, the tracings of the two specimens indicate the same species — not however, that of the more northern and early accessible eastern coastal region, but that of the interior and extreme southerly and later accessible eastern coast.

About the beginning of this century the two more common species of Florida were clearly distinguished, as already mentioned on a preceding page, both as to the specific characters and to the geographic ranges.[1] The one of more northern distribution was assigned to hammocks of middle peninsular Florida, particularly on the eastern coast; the other was assigned to flat-woods—pine-woods—east coast of Florida, below New River, Broward County.

It is of particular interest that the two perfectly distinct habitats for *Zamia* were recorded by Bartram and Baldwin respectively, about the beginning of the nineteenth century. The two species, however, were referred to under the botanical name *Zamia pumila*—a species inhabiting Hispaniola, and perhaps also some of the other West Indies.

William Bartram[2], writing of experiences and objects in the

---

[1] Herbert John Webber, Bulletin of the U. S. Department of Agriculture 1: 81. 1901.

[2] William Bartram was born 9 February 1739, at the botanic garden of his father, John Bartram, at Kingsessing, near (now in) Philadelphia. He had the advantage of a better education than his father, and was an artist of considerable ability. He was a clerk in Philadelphia for a few years, and then a merchant in Carolina, but he was more interested in botany than business. In 1765 he joined his father in exploration in Florida, and when his father returned home in 1766 he remained as a settler on the St. John's; but the next year he returned to Kingsessing. From 1773 to 1778 he was engaged in botanical travels in the Carolinas, Georgia, and Florida, of which

Lake George region of Florida, records[1]: "The *Zamia pumila*, the *Erythryna corallodendrum*, and the *Cactus opuntia* grow here in great abundance and perfection. The first grows in open *pine forests*, in tufts or clumps, a large conical strobile disclosing its large coral red fruit which appears singularly beautiful amidst the deep green fern-like pinnated leaves." The italics in the case of "*pine forests*" are ours. Likewise in the case of "*hammock*" in the following quotation.

William Baldwin[2] writing of experiences and objects along the lagoons south of the mouth of the Saint Johns River, records:[3] "Next morning, crossing the Inlet at the southern point of Penon Island, we ascended upper Matanza river,—which running south 10 or 12 miles close to the shore, originates by several heads in the swamps, a little westerly. Tracing the eastern branch, we landed early in the afternoon at the plantation of a Mr. Hernandez. Here, in a thin sandy *hammock* of small Live Oaks, Cabbage and Saw Palmettoes, I had the gratification to find the "Wild Sago," or Coontia, of the Seminoles,—and to assign it its place in the sexual system; Dioecia, Polyandria: natural order, Palmae. I have no books with me to refer to; but it is probably a new genus,—approaching very closely in habit to

an account was published in book form in 1791. The rest of his life was spent in scientific study at the garden at Kingsessing, in the homes of the owners of the garden—at first his brother John, later Colonel Carr—and it was there that he died, 22 July 1823.—John Hendley Barnhart.

[1] Travels through North and South Carolina, Georgia, East and West Florida 160. 1792.

[2] William Baldwin was born in Newlin township, Chester County, Pennsylvania, 29 March 1779. After practicing medicine for several years, and taking a partial course in this subject at the University of Pennsylvania, he shipped as surgeon in a merchant vessel sailing from Philadelphia in 1805, by way of Antwerp, for China. Returning the following year, he completed his medical course, receiving his doctor's degree in 1807. He practiced for four years at Wilmington, Delaware, removing to Georgia in 1811. In the following year he became surgeon in the United States Navy, and was stationed at St. Mary's for two years and a half, and for two years at Savannah. From March to May 1817, he visited Florida, devoting most of his time to botanical exploration. Later in the same year, as a naval surgeon, he accompanied a United States mission to Buenos Aires and other South American ports. Upon his return the following year he lived at Wilmington, Delaware, until he was appointed surgeon and botanist to Long's expedition up the Missouri. His health, always delicate, broke down soon after the expedition was under way, and he died at Franklin, Missouri, 1 September 1819. Many of his letters were published in 1843, by his friend Darlington, under the title "Reliquiae Baldwinianae."— J. H. B.

[3] Reliquiae Baldwinianae 225, 226. 1843.

the real Sago family (Cycas). At supper, I had the pleasure to eat the bread prepared from the large tuberous root of this plant. In the late times of difficulty many negroes, and others, were prevented from perishing with hunger by having recourse to it; and the slaves on this plantation now save half their allowance, in consequence of using it. I have no hesitation in saying that it will be found among the most important of our Esculentia. I believe I have already given you a hint of this plant. At some future period, I will give you more particular information." "Bowlegs, the grandson of Bartram's 'Long Warrior,' says, that 'Coontia' signifies Bread plant." (This proved to be the *Zamia pumila*. See subsequent letter, of May 27.) The above letter was written on May 15, 1817, from Tomoko, which is the old name for the site of the modern Daytona.

Later, on May 27, 1817, in writing from Saint Marys, Baldwin records: "I now find that my Coontia, or "Wild Sago," is nothing more or less than *Zamia pumila*."

While perusing some of the letters in the volumes of selected correspondence of Linnaeus and other naturalists, after the foregoing matter was written, we stumbled on the following paragraphs written by Alexander Garden[1] from Charleston, South arolina, to Linnaeus in Sweden, and to John Ellis[2] in London,

[1] Alexander Garden was born 20 January 1730, at or near Edinburgh, Scotland. Educated at Aberdeen and at Edinburgh, where he took his medical degree, he settled in South Carolina in 1752, and soon afterward made his home at Charleston, where he practiced medicine and studied the flora and fauna of the colony for nearly thirty years. During a northern journey in 1754 he became personally acquainted with Bartram and Colden and in the following year opened correspondence with Linnaeus. He was also for years a correspondent of Ellis, Collinson, Gronovius, and other botanists. Ellis named in his honor the genus Gardenia, known to all lovers of flowers. When the Revolution broke out, his only son joined the American troops, and was never forgiven, for the father was an ardent loyalist. In 1783, with his wife and daughters, he left South Carolina, and spent his remaining years in London, where he died 15 April 1791.— J. H. B.

[2] John Ellis was born in Ireland, probably in Dublin, about 1710. He became a merchant in London, and amassed a considerable fortune. In 1751 he became interested in the study of marine algae and other forms of aquatic life, especially what were then called "corallines," which he soon realized were some of them animals and others plants. Later he studied fungi and flowering plants. He was appointed King's agent for West Florida in 1764, and for Dominica in 1770, and imported to England many scientific specimens of various sorts, especially seeds of useful plants. Among his correspondents were Linnaeus, and Dr. Alexander Garden of Charleston, South Carolina. He died at London, 15 October 1776.—J. H. B.

England: "The doctor [Andrew Turnbull] carries home some packages of East Florida plants, which you will see. I shall be very glad to know what you make of John Bartram's[1] Tallow tree, and what you call that herb whose leaves look like the Fern Osmund Royal, while its seeds are large red berries in a cone, somewhat resembling the Magnolia in appearance. I shall be glad to know what you call these two."[2] (Written February 2, 1767.)

The specimens are said to have reached Princess of Wales Garden at Kew, England, in 1768. There they came to the notice of William Aiton.[3] The interest once aroused continued for several years, as will be seen by the following quotations.

A few years later Alexander Garden writing to Linnaeus, said: "In the same parcel with the fishes, I have sent, for your examination, a Florida plant, unknown to me, which should seem to belong to Gynandria. Its appearance is handsome enough, especially in autumn, when the woods are much ornamented with its beautiful fruit and seeds. This specimen shows the flowers in a head, as they at first appear; but they are soon succeeded by a cone-like pericarp, nearly of the same figure as that of a *Magnolia*. At length the capsules burst into two parts, displaying the large shapeless seeds, which turn red as they ripen,

[1] John Bartram was born 23 March 1699, at Marple, near Darby, Pennsylvania. He had been a farmer for several years before he became interested in botany, and was then for the most part, and of necessity, as he knew no one else interested in the science, self-taught. He traveled widely, from New York to Florida, and corresponded with various eminent naturalists abroad, especially with Collinson of London, who was, like himself, a member of the Society of Friends. About 1729 Bartram established at Kingsessing, then a surburb but now a part of the city of Philadelphia, the first botanic garden in America. For nearly fifty years an almost continuous stream of American seeds and plants, sent by Bartram, poured into the gardens of Europe. In 1765 he was appointed King's Botanist, and the modest salary accompanying this appointment enabled him, in spite of advancing age, to extend his scientific travels. In that year, and the next, accompanied by his son, he explored the St. John's River, Florida, and several of its tributaries. He died at Kingsessing, 22 September 1777.—J. H. B.

[2] Correspondence of Linnaeus and Other Naturalists 1: 552. 1821.

[3] William Aiton was born in 1731, at Hamilton, Lanarkshire, Scotland. As a young man he went to London, and was employed by Philip Miller at the Chelsea Physic Garden. In 1759 he was engaged by Augusta, Princess of Wales, to establish at Kew, her residence, a "physic" or botanic garden, and it is from this event that the present Royal Gardens, as a scientific institution, date. In 1789, he published his "Hortus kewensis," and remained superintendent at Kew until his death 2 February 1793.—J. H. B.

and attract notice from a considerable distance. The height of the plant is 18 inches, rarely more. I am extremely desirous of knowing what it is, but dare not dissect this only specimen that has been brought me, which will be better in your hands. Formerly I have been in the way of seeing plenty of both fruit and seeds."[1] (Written June 20, 1771.)

Several years later Alexander Garden in writing to John Ellis again referred to the Bartram plant as follows: "I have taken the liberty of enclosing a letter for Linnaeus under this cover, and I beg you will forward it to him as soon as possible. It relates chiefly to an East Florida plant, of which I formerly sent you a fruit with the seeds therein. I have now carefully examined it, and you will see the characters, which are somewhat like the characters of the Zamia, but yet I think different. I have sent specimens of this plant, and of all parts of it, in a bundle directed to you, and given to Capt. William White's particular care. You will greatly oblige me if you will desire him to send these things to your house, and then forward them to Linnaeus."[2] (Written May 15, 1773.)

Then, in a letter to Linnaeus, referred to in the preceding quotation, Alexander Garden wrote:

"I am always glad to be employed in your service, of which I trust you are by this time well assured. That I might give a proof of this, on the receipt of your letter dated Upsal, the 16th of January, 1772, I took measures to procure from East Florida, where it grows wild, some specimens of your *Zamia*. I wrote to the Governor of that province, a great promoter of Natural History, and my particular friend, requesting him to send me some, carefully gathered, and well preserved, by the first opportunity. According to his usual kindness, he sent me what I wanted. as soon as the season of the year would permit, and from these specimens I have made out the following particulars and characters.

"You mention in your letter, that the pollen of this plant is naked, on the under surface of each scale of the cone (or catkin). But the most careful examinations, under a microscope, have satisfied me that this is by no means the case. I was much afraid of committing a mistake, and leading others into error, and therefore submitted several scales to repeated investigation, always with the same success, for I saw no difference between them. The pollen is evidently contained in bivalve, elastic capsules (anthers). It can scarcely therefore be referred, as you judge, to the Fern tribe, nor dare I assert it to be a *Zamia*. The construction of the female *spadix*, and of the pericarp, is very

[1] Correspondence of Linnaeus and Other Naturalists 1: 336-337. 1821.
[2] Correspondence of Linnaeus and Other Naturalists 1: 598-599. 1821.

Plate 259

Underground stems of *Zamia integrifolia* from the sand-filled cavities in the honeycombed floor of the pinelands. The slab in plate is over one foot wide. The stem of zamia is usually described as simple in this species, at least, branched stems are not rare. Great irregularity in form and branching obtains. They produce a percentage of a beautiful white starch which is freed from the wood and other impurities by maceration and washing. The spent material, shown in a pile behind the slab, is locally used as fertilizer. The starch is always powdery as the grains will not adhere in lumps.

singular. The peltate heads of the proper perianths are externally so closely united, that they can hardly be pulled asunder without tearing. It is therefore scarcely to be understood how the pollen of the anthers can insinuate itself so as to fertilize the germens. For the perianths never begin to separate before the germens become swelled, exhibiting manifest signs of impregnation being already accomplished. The vacant internal space indeed, between the partial stalks of the perianths, affords the germens and styles full liberty to grow; but the very close union of the shields prevents any access of external bodies, or even of the most subtile vapour.

"Being anxious to know more of this plant and its history, I have put the following questions to my friend the governor, by letter. Is there any *Spatha* or not? Are the male and female catkins always on the same plant, or from separate roots, and is there no variation in this respect? I have some suspicion that fecundation may take place sometimes in one way, sometimes in another. Do any birds, and of what kinds, feed on the seeds? Does the plant afford nourishment to any other animals? I am in daily expectation of an answer.

"It remains with you to determine the genus of this plant, and that you may not want materials to foim your judgment upon, I now send you dried specimens of the male and female catkins, as well as of the seed-vessel or cone, containing ripe seeds."[1] (Written May 15, 1773.)

A detailed description of the staminate and ovulate cones is appended to the letter, under the heading of "Characters of the dwarf palm from East Florida." Our Zamia was evidently quite a puzzle to the early botanists, for, we find them at one time or another, referring to the plant as, a *fern*, a *cycad*, a *palm!*

The political conditions and the approaching American Revolution were fatal to Alexander Garden's plans and correspondence, and nothing further appears with reference to *Zamia*.

Twenty-one years passed after the introduction of the John Bartram plant into Kew before the gardener there described and published it under the name *Zamia integrifolia*. Thus we have documentary evidence as to the collector of that species, and its origin was doubtless in the watershed of the lower St. Johns River, perhaps in the Lake George region, where William Bartram also found it as stated in a foregoing quotation.

Now, to pass to the second described Florida zamia. The discovery of documentary evidence concerning the origin and dis-

---

[1] Correspondence of Linnaeus and Other Naturalists 1: 338-340. 1821.

coverer of *Zamia floridana* seemed even more hopeless than it did in the case of *Z. integrifolia*. However, we were not wanting for circumstantial evidence. In publishing *Zamia floridana*, Alphonse de Candolle[1] merely says, "E Florida sub nom. evidenter erroneo Z. integrifolia a cl. A. Gray, a. 1839." Now as A. Gray[2] had sailed for Europe in the fall of 1838, he had, evidently, taken specimens of this zamia with him and given them to Augustin de Candolle.[3] However, their source was still a

[1] Alphonse Louis Pierre Pyramus de Candolle, son of Augustin Pyramus de Candolle, was born at Paris, France, 27 October 1806; the family removed to Montpellier two years later. In 1814, however, he was sent to school at Geneva, Switzerland, and his parents later made their home at the same place. He studied law, receiving his degree in 1829, but he had previously done much work in botany, and thereafter devoted himself entirely to that science. He continued and completed the "Prodromus" begun by his father, and was the author of many important monographs, and of books on geographic botany, on the history of sciences, and on the origin of cultivated plants. He died at Geneva, 4 April 1893, after more than sixty years of scientific activity.—J. H. B.

[2] Asa Gray was born 18 November 1810, at Sauquoit, in Paris township, Oneida County, N. Y. While a student of medicine at the medical college at Fairfield, Herkimer County, he became interested in botany, and although he took his degree in 1831, he never practiced medicine. During the next few years he taught and lectured on mineralogy and botany at Albany, Clinton, Fairfield, and Utica, and in New York City, where he first went as an assistant to Torrey, and later became associated with him in the publication of their "Flora of North America." In 1842, he became a professor at Harvard, and it was here that he accomplished, during his forty-six years of service, the vast amount of careful work that secured him undisputed recognition as the foremost American botanist. He died at Cambridge, Massachusetts, 30 January 1888.—J. H. B.

[3] Augustin Pyramus de Candolle was born 4 February 1778, in Geneva, Switzerland. He was educated in his native city, and began the study of law there, but before he was twenty years old he went to Paris to pursue natural history studies. Coming under the influence of Lamarck, Desfontaines and other botanists, he decided to devote himself exclusively to the study of plants. After about ten years of study in Paris, during which he began the copious contributions to botanical literature which soon made him famous, he became in 1807 professor at the university of Montpellier and soon afterward director of the botanic garden there. In 1816 he returned to Geneva as professor in the university, and there rem ined until his death, 9 September 1841. He is best known to botanists, perhaps, by the "Prodromus" started by him in 1824, and completed by his son in 1873, but this great work was largely a compilation, while de Candolle published many important monographs and several monumental volumes devoted to the results of his original investigations.— J. H. B.

mystery. Now, in the acknowledgments of assistance in the preface to Torrey and Gray's "Flora" we find recorded: "From Middle Florida . . ; while from Southern and Eastern Florida we have received interesting collections from Dr. Leavenworth,[1] Dr. Burrows, Dr. Hulse, and Lieut. Alden[2] of the United States Army." All things considered, these facts and circumstances seemed to focus on one or more of the four army officers just mentioned as the possible collectors of the plant in question.

A scanning of the unpublished letters from these men to John Torrey[3] brought out the following, which is contained in some

[1] Melines Conklin Leavenworth was born 15 January 1796, at Waterbury, Connecticut. He studied medicine at Yale, receiving his degree in 1817, and practicing his profession at Catawba and Augusta, Georgia. From 1833 to 1840 he was a surgeon in the United States Army; stationed at first at Fort Towson, "Arkansas" (eastern Oklahoma); then at Fort Jesup, in western central Louisiana, doing some collecting in eastern Texas; from May to November, 1838, at Fort Micanopy, in north central peninsular Florida; then for a few weeks thirty miles east of Tallahassee; and during the first half of 1839, at Fort Frank Brooke, four miles from the mouth of the Steinhatchee River. During the remainder of his service he was stationed at Fort Gratiot, Michigan, but he resigned from the army in 1840, and retired to his old home at Waterbury, Connecticut. In 1862, although 66 years old, he enlisted as surgeon in the 12th regiment of Connecticut militia, and served with them in Louisiana until death claimed him. During all his travels he devoted his spare time to the collection of plants, which he sent to Short, Torrey, and others. He died near New Orleans, Louisiana, 16 November 1862.—J. H. B.

[2] Bradford Ripley Alden was born 6 May 1811, at Meadville, Pennsylvania. He graduated at West Point in 1831, and served in Florida as a lieutenant in the 4th Infantry, 1832-33, at first at Fort Brooke (Tampa), but most of the time at Fort King (Ocala); it was here that he made a considerable collection of plants for his friend Dr. Torrey. He was afterward instructor at West Point for seven years (1833-40) and commandant there for seven more (1845-52); wounded in service, he resigned from the army in 1853. He never fully recovered his health during the subsequent seventeen years of his life, and died at Newport, Rhode Island, 10 September, 1870.—J. H. B.

[3] John Torrey was born in New York City, 15 August 1796. As a boy he became interested in botany and chemistry. When he was only twenty years old, and a student of medicine, he was one of the group of young men, who, under the leadership of Professor Mitchill, organized the Lyceum of Natural History, now the New York Academy of Sciences, and within a year, as one of the members of a committee appointed by the Lyceum, prepared a catalogue of plants known to occur within thirty miles of New York City. Upon receiving his degree, in 1818, he entered upon the practice of medicine in New York City; from 1824 to 1827, he was professor of chemistry

letters from G. W. Hulse[1]. Writing from Fort Brooke (now Tampa), East Florida, under date of June 10, 1837, he said:

"I should apologize for the small offering that I am about to send you, but I always dislike making excuses. The few seeds that I send are from the plant that is indigenous in Florida and called the *Arrow Root*. Perhaps they will vegetate with you and give you opportunity of saving the plant itself. It grows abundantly near Cape Florida, the mouth of the Withlacoochee and Sawanee Rivers. Quite extensive establishments were fitted up at the Cape before the war for the purpose of manufacturing the root into starch (or rather flour) which is said to be a very simple process, merely grinding or grating the roots, then washing, precipitating and drying. The article when prepared is said to be equally as good as the *Maranta Arundinacea*, though this I believe is quite a different plant. It is said that it grows best on sandy light soil. I have seen some roots that would weigh several pounds.

I send you also a sample of Seminole Flour which was taken from the stores of the enemy at the Big Cypress near Tohope-

and mineralogy at West Point; from 1827 to 1873, professor of chemistry at the College of Physicians and Surgeons (now the Medical Department of Columbia University); for some years also concurrently professor of chemistry at Princeton; botanist of the geological survey of New York State; and from 1857 to 1873, United States Assayer at the New York office. In spite of the fact that his professional work was all in medicine, chemistry, and mineralogy, his first interest was always in botany, to which he devoted all the time he could find for the purpose for more than fifty years, publishing numerous books and papers, and gaining a world-wide reputation in this science. He was the first president of the Torrey Botanical Club, which was named in his honor. He died in New York City, 10 March 1873.—J. H. B.

[1] Gilbert White Hulse was born 12 March 1807, in BloomingGrove township, near Washingtonville, Orange County, New York. He studied medicine at the College of Physicians and Surgeons in New York City, receiving his degree in 1835,. He immediately entered the medical service of the United States Army, being stationed at Tampa Bay from February to April, 1836, in Arkansas during the summer and fall of the same year, returning to Fort Brooke, Florida. In 1837 he went again to Fort Gibson, Arkansas, returning to Fort Brooke in January, 1838. He was at Tallahassee in March, and before the middle of the year 1838 had settled as a medical practitioner at Rodney, Mississippi. Later he was a planter in Louisiana, and about 1850 he visited northern California, where he collected plants for Torrey as he had done in Florida and Mississippi. Throughout the Civil War he was a surgeon in the Confederate Army, and at its close., his property gone, he returned north, making his home at first with his sister, Mrs. Moffatt, at Rockford, Illinois, and later with her daughter, Mrs. Knapp, at Auburn, New York, where he died 13 November 1883.—J. H. B.

Ka-Lija Lake on the 28th of January. It is made, I believe, from a plant which is a species of Briar very abundant in the south and in Florida, grows best in the rich Hammock land, and if we should judge its use by the establishments that we have seen for preparing it, we should say it was a staple article with the Seminoles. It is called by them *Conti Chatee* or Red Root whilst the arrow root which is also much used by them is called *Conti Hateka* or White Root."

A letter written early in the following year from Tallahassee, March 7th, 1838, said: "I owe you many apologies for having delayed so long to acknowledge your favour of the 22nd of August last. It did not reach me however, till the 3rd of January when I was at New Orleans on my return from Fort Gibson to Fort Brooke E. F.

"It affords me real satisfaction to say to you, I have according to your request been able to collect and forward you some specimens of the plant *Zamia integrifolia.*

"They were sent by the Brig Wm. Penn which sailed from St. Marks for New York on the 5th inst. In the collection I have been careful to take an assortment of both sexes or males and females having the strobiles so well developed that in case they reach you without injury (on board of the vessel) you will be able to see much that is peculiar and interesting in the botanical character of the plant. They are from two different localities: those with large, long, roots are from the sandy pine woods; those from the sea shore (mouth of Withlacoochee) grew very near the salt water and upon the beds of oyster shells and in consequence have small roots.

"I regret that I have not been able to furnish you with some more of the ripe fruit or seeds. The few that I sent you with the other (last) parcel are all that I have seen. The fruit is an oblong shape and when ripe nearly the size of a grain of gourd seed corn."

Under date of July 7, 1838, Dr. Hulse wrote from Rodney, Mississippi, as follows:

"My last letter to you was from Tallahassee (Fla.) to which I have received no answer. It was dated about the middle of March, a little previous to which time I sent you by a vessel (Brig Wm. Penn) from St. Marks to N. Y. a qu ntity of the *Zamia* put up as you directed."

Referring to the "Perrine Grant" a reservation in southern Florida, intended for growing useful plants from the tropics, he continues:

"It is my opinion however, that the *Zamia integrifolia* will be found a no less important plant than many of those proposed to be introduced." This statement is partic larly interesting since the pinelands of the "Perrine Grant" naturally abounded in zamia, and the supply has been generously drawn on for several generations.

There is scarcely any doubt that some of the material of zamia referred to in the preceding letters of Dr. Hulse reached New York before Asa Gray sailed for Europe. It was doubtless some of that material that Dr. Gray took to Geneva. At any rate, the tracing of the type specimen of *Zamia floridana* indicates the plant that is common in the pine woods east of Fort Brooke where the writer has collected it.

One of the more interesting statements in one of the above letters is that referring to the two habitats of the *Zamia* in question. As just stated, the pineland plant represents *Z. floridana*. The one growing in the oyster shells at the mouth of the Withlacoochee may represent another species.

There is a sheet in the Torrey herbarium with two rather poor leaves representing two species of zamia, and the record "Florida Dr. Hulse & Dr. Burroughs." These leaves may be from the two kinds of plants with dissimilar "roots" mentioned in Dr. Hulse's letter of March 7, 1838. Whether the leaves came from the plants when they were sent to Dr. Torrey, or whether they were from the "roots" grown under glass at New York we shall never know. One leaf has narrow, few-veined leaflets, the other has them broader and many-veined.

What part Dr. Burrows played in this *Zamia* puzzle we do not know, as there are none of his letters preserved in the unpublished Torrey correspondence at the Garden and he is not referred to in the Hulse letters. Of course, it is possible that the two leaves were sent in independently. If the leaves do not represent the plants referred to in above cited letter, Dr. Hulse may have sent the one with narrow leaflets to Dr. Torrey and Dr. Burrows the one with broad leaflets, or vice versa.

William Bartram found zamia on the western side of the State as well as on the eastern. While wandering in the wilderness east of the Suwannee River he records that he "had an opportunity this day of collecting a variety of specimens and seeds of vegetables, . . ., particularly Sophora, Cistus, . . . Zamia, . . ."[1] The writer found zamia in this region several years ago.

[1] William Bartram, Travels through North and South Carolina, Georgia East and West Florida. 246. 1792.

Zamia has been rather an elusive plant in Florida. For many years only two general localities were associated in literature with the geographic distribution of the two generally recognized species, whereas the plants are widely scattered over the peninsula.

The two species just referred to are quite distinct. The one, with broad and many-veined leaflets, occurs most abundantly in the hammock-belt of the Halifax River on the upper eastern coast. Curiously enough this was the Florida cycad field most easy of access in the early days, but that *Zamia* was referred to only by William Baldwin, as *Zamia pumila*, and to this day the species has not received a name that can be retained for it, as it differs from the true *Zamia pumila* L. of the West Indies.

The other species, with narrow and few-veined leaflets, the one that figured prominently in the early botanical literature and correspondence, a plant of the pinelands, was discovered not far from the Halifax River region, just across the watershed separating the Saint Johns River from the eastern coast, and named *Zamia integrifolia*, by William Aiton in 1789. It was discovered independently some fifty years later evidently near the western coast, directly west or southwest of the Halifax River region, and named *Zamia floridana*, by Alphonse De Candolle, in 1868. The most interesting fact in connection with the geographic distribution of this species is that its greatest abundance is at the very opposite extremity of the peninsula from the place of its discovery. The natural growth of zamia on the Everglade Keys is phenomenal—perhaps the most luxuriant in existence. There it was and is so abundant that starch or flour has been manufactured from it for ages under the name of "Conti-hateka" or "Coontie" by the red man, aborigines and Seminoles, and "Florida arrow-root" by the white man. It is sometimes recorded that the plant was there cultivated, but all our information is to the contrary. The natural abundance evidently deceived some observers. This exceptional abundance was observed long ago, for on some of the early maps the region which we now know as the Everglade Keys bore the significant legend "Koontee or Hunting Grounds!"

The writer has been gathering evidence concerning the geographic distribution of *Zamia* in Florida for several years.

The evidence in hand indicates major phytogeographic regions: I. For *Zamia integrifolia* Ait., discovered by John Bartram, —the pinelands as noted above locally throughout the peninsula, with the plants most abundant near the northern and southern ends, with a northward extreme at Perry in Taylor County and a southern outlying extreme on Big Pine Key. II. For *Zamia umbrosa* Small[1], discovered by William Baldwin, as noted above, —the temperate hammocks of the upper eastern coastal region with outlying extensions in the Saint Johns water-shed and the Ocklawaha water-shed. III. For *Zamia media* Jacq., of the West Indies, discovered in Florida in 1917, by C. A. Mosier, John De Winkeler and the writer,—the tropical and semitropical hammocks of the lower eastern coastal region.[2] IV. For an undertermined species of *Zamia*—the tropical everglade prairie hammocks of the Cape Sable region.

Rumored more northern localities for *Zamia integrifolia* and *Z. umbrosa* await verification. The former species has been reported as growing south of De Funiak Springs, western Florida and the latter north of Saint Augustine, eastern Florida.

Thus, one after another some of the mysteries have been solved. The one, however, connected with the general distribution of these plants over the state is unsoluable. Whether, after they were generally established, subsequent to the immigration of their ancestors from the West Indies, they have increased in range or abundance, naturally, or through the agency of the aborigines, as a cultivated crop, or have decreased through natural agencies or through the abuse of the supply by the aborigines, we shall never know. However, at present it is clear that the plants are most abundant at the sites of the former places of settlement or activity of the aborigines.

[1] **Zamia umbrosa** n. sp. Plant with arching, dark-green leaves: leaflets typically numerous, the blades narrowly spatulate at least broadened upward, 20-30-veined, finely several-toothed at the apex: mature, ovulate cones ellipsoid or cylindric, 1-2 dm. long, or rarely smaller, scarcely umbonate.—Hammocks, shell-middens, and sand-dunes, northeastern peninsular Florida.—Type specimen from Hammock, between Volusia and Ocean City, Florida, J. K. Small & J. B. DeWinkeler, May 4, 1821.

[2] Journal of the N. Y. Bot. Gard. **18:** 102. 1917.

Zamias are ornamental as well as esculent plants. In Florida they are commonly cultivated in gardens, not only in the localities where they grow naturally, but at distant points. They decorate the front yards of both humble and pretentious houses, planted either in clumps or as hedges. When the Everglade Keys were settled, every one had fine plants of *Zamia integrifolia* in his front and back yards. These usually grew so plentifully that it was a case of eliminating instead of introducing the plants.

The plant thrives in cultivation. Garden specimens often surpass in luxuriance any seen in their natural habitats. Thus we have found such specimens of *Zamia integrifolia* in Apalachicola and in Perry in northern Florida, and likewise in towns down through the peninsula. *Zamia umbrosa* is in cultivation west of its natural range in Gainesville and in Ocala. Northward of its geographic range it may be seen in abundance in the gardens of Saint Augustine and in Jacksonville, and even in the little, remote, but old, settlement of Mayport at the mouth of the Saint Johns River.

Numerous inquiries in Saint Augustine and in Jacksonville brought out the fact that the cultivated plants were brought originally from the Halifax River region, but long before the present owners or occupants of the premises were there.

The individual plants grow rapidly from seed. Seedlings may become mature enough in two or three years to bear cones, even under unfavorable conditions. They, too, seem to be long-lived, and several centuries may not be too great an estimate. An evidently old plant, moved to a garden in Pasco County, Florida, over thirty-five years ago thrived and was sent to the to the New York Botanical Garden within the past few years but unfortunately, was frozen en route.

Under natural conditions the plants are scattered broadcast by the dispersal of their seeds by various anima's; but vegetative propagating may be easily accomplished. The curious-looking caudex is evidently a much abbreviated and condensed plant axis, for it may be sliced into numerous wafers, each of which will promptly develop into a new individual if planted. It seems strange that in spite of these numerous nascent buds, the caudex of our zamias is either simple or very sparingly branched, both in the wild and in cultivation.

JOHN K. SMALL.

THE NEW HORTICULTURAL GARDENS ENTRANCE AND FENCING ON THE SOUTHERN BOULEVARD.

WITH PLATE 260

The fencing of the southern part of the garden reservation, along the Southern Boulevard and Pelham Parkway, made necessary by largely increasing numbers of visitors and the protection of plantations and of natural features was commenced last autumn, and is being continued. The fence is of the same design as that built in previous years along the property line of Fordham University on the southern side of the garden and along the Bronx Boulevard on our eastern boundary; it is composed of a low rubble-stone wall, built on a concrete foundation and topped by a coping, with square piers of Yonkers granite about twenty-five feet apart surmounted by caps, and with bays of iron railing about seven feet high. The new fencing differs from the old in that the coping and pier caps are of artificial cast stone instead of granite.

The new work was commenced at the Southern Boulevard Entrance and about five hundred feet of fencing has now been completed, including a path entrance to the Horticultural Gardens. This entrance consists of two cut cast stone piers seventeen feet apart. Work on the continuation of this fence southward is in progress and it is hoped to complete about two hundred and fifty feet more of it this season. Over this length and beyond toward Pelham Avenue the fence will rest for the most part on a natural rock foundation.

The cut cast stone piers at the Horticultural Gardens Entrance and about three hundred feet of the fence were built by the expenditure of a bequest of $5,000 by Mrs. Mary J. Kingsland, appropriated by the Board of Managers for this purpose. A bronze tablet, affixed to one of the piers bears the following inscription.

This Entrance and
fence were built
1920-1921
through a bequest of
Mary J. Kingsland
New York Botanical Garden

The New Horticultural Garden Entrance and Fencing on the Southern Boulevard

The tablet was unveiled during the Spring inspection of Grounds, Buildings and Collections on May 6th, 1921, by Mrs. Theron G. Strong, Secretary of the Women's Auxiliary.

The rest of the fence completed has been built by expenditures from the income of the Russell Sage and Margaret Olivia Sage Memorial Fund, and the work in progress is being financed from this income.

THE 1921 DAHLIA BORDER

The 1921 dahlia border is at the date of writing (July 15) well established and, after the incipient drought of the first three weeks of June, has been receiving a supply of rainwater sufficient to meet its needs. About sixty of the nearly five hundred varieties that were carried through the winter were rejected, not that they were really poor varieties, but because among the newer creations of the last two or three years there seemed to be better ones, to the growing of which our limited space might better be devoted. The missing numbers have been more than made up by contributions from various friends and patrons of the border, among whom may be mentioned Judge Josiah T. Marean, Green's Farms, Conn.; Mr. J. J. Broomall, Eagle Rock, California; Mr. F. P. Quinby, White Plains, N. Y.; Mrs. Charles H. Stout, Short Hills, N. J.; Mr. W. J. Matheson, Huntington, N. Y.; Mills & Co., Mamaroneck, N. Y.; Miss Emily Slocombe, New Haven, Conn.; Mr. Alfred E. Doty, New Haven, Conn.; Mr. C. Louis Alling, West Haven, Conn.; Mr. Alt F. Clark, Netcong, N. J.; Mr. Emmett Dove, Rockville, Md.; Mrs. A. F. Story, Brockton, Mass.; Capt. J. R. Howell, Bayshore, N. Y.; Mrs. S. T. Cushing, Islip, N. Y.; Mr. R. Vincent, Jr. & Sons, White Marsh, Md.; Dahliadel Nurseries, Vineland, N. J.; and The Dahlia Farm, East Moriches, N. Y. The dahlia border this year includes 502 varieties, represented by 824 plants. It is believed that the addition of numerous tested novelties from such well-known originators as Broomall, Marean, Slocombe, and Stout, together with other promising new varieties that have been perhaps less thoroughly tried out, will serve to maintain the interest of the New York public in the Garden's dahlia border and will also do much to assist the public to keep in touch with the most recent perfections in the development of this popular flower.

MARSHALL A. HOWE

AUTUMN LECTURES

Illustrated lectures free to the public will be delivered in the lecture hall of the museum building on Saturday and Sunday afternoons at four o'clock during September and October as outlined below. The following lecturers are to be with us for the first time: Mr. John Dunbar, of the Rochester Park System; Dr. Edgar T. Wherry, of Washington; Prof. T. Gilbert Pearson, President of the Audubon Society; Dr. H. M. Denslow; and Miss Frances B. Johnston.

During November, lectures will be given in Conservatory Range 2 on Sunday afternoons at a quarter past three o'clock. Mr. Arthur Herrington will open the series with a talk on "Chrysanthemums," illustrated with a collection of living plants and cut flowers.

REGULAR COURSE

Sept. 3. How to Grow Rhododendrons. Mr. John Dunbar

Sept. 4. The Classification of Plants. Dr. W. A. Murrill
(With Museum Demonstration)

Sept. 10. How to Grow Wild Flowers. Dr. E. T. Wherry

Sept. 11. English Gardens. Miss Hilda Loines

Sept. 17. Some Interesting Plants of Our Local Flora.
Dr. F. W. Pennell

Sept. 18. The Fight for American Bird Protection.
Prof. T. G. Pearson

Sept. 24. Dahlias and Their Culture. Dr. M. A. Howe
(Exhibition of Dahlias, September 24 and 25)

Sept. 25. Our American Gardens. Miss F. B. Johnston

Oct. 1. The Uses of Plants. Dr. W. A. Murrill
(With Museum-Demonstration)

Oct. 2. A Popular Talk on Mushrooms. Dr. W. A. Murrill

Oct. 8. Loco Weeds. Dr. Arthur Hollick

Oct. 9. Health and Disease in Plants.   Dr. A. H. Graves
Oct. 15. Autumn Coloration.   Dr. A. B. Stout
Oct. 16. Rice, the Most Important Food Plant.
                                           Dr. H. A. Gleason
Oct. 22. Our Local Orchids.   Dr. H. M. Denslow
Oct. 23. Gathering Wild Flowers in the Tyrolean Mountains.
                                           Dr. W. A. Murrill
Oct. 29. Florida Vegetation.   Dr. J. H. Barnhart
Oct. 30. The Influence of Climate on Evolution.
                                           Dr. W. A. Murrill

GREENHOUSE LECTURES

Nov. 6. Chrysanthemums. Mr. Arthur Herrington
Nov. 13. Tropical Vines. Mr. K. R. Boynton
Nov. 20. Cycads. Dr. W. A. Murrill
Nov. 27. Variegated Plants. Dr. A. B. Stout

Conservatory Range 2 is situated at the eastern side of the Botanical Garden, north of the Allerton Avenue Entrance. It is most conveniently reached from the Allerton Avenue Station on the White Plains Extension of the Subway from East 180th Street. Visitors coming by train to Botanical Garden Station should inquire at the Museum Building.

NOTES, NEWS AND COMMENT

On the afternoon of July 13, a group of about sixty students from the Columbia University Summer Session visited the Garden under the leadership of Mr. L. A. Crawford, assistant to the director of the Summer Session. The party was met at the Elevated Railway Station by members of the Garden Staff and escorted through Conservatory Range I, the Herbaceous Grounds, the Rock Garden, the Iris Garden, the Hemlock Grove, and the Museum Building.

By exchange of duplicate specimens with the Smithsonian Institution, the herbarium has recently been enriched by over 1200 specimens collected in Haiti last year by Mr. E. C. Leonard. This is one of the largest collections ever made in Haiti, and it contains specimens of a good many species not previously represented in our collections and many other specimens of rare trees and shrubs endemic in Haiti. It is very valuable to us at the present time in connection with our studies of the flora of Cuba and of that of Porto Rico.

Meteorology for June: The total precipitation for the month was 3.02 inches. The maximum temperatures recorded for each week were as follows: 84° on the 4th, 88° on the 12th, 87° on the 13th, and *98° on the 22nd*. The mimimum temperatures were *45½° on the 3rd*, 50° on the 7th, 53° on the 16th and 60° on the 21st.

ACCESSIONS

PLANTS AND SEEDS

60 plants for Fern Garden. (Given by Mr. C. A. Weatherby.)
18 plants for Iris Garden. (Given by Rev. C. H. Demetrio.)
13 plants for Iris Garden. (Given by Miss Grace Sturtevant.)
4 plants for Iris Garden. (Given by Mrs. D. A. Filler.)
3 plants for Iris Garden. (Given by Mr. S. G. Harris.)
20 plants for Iris Garden. (Given by Mr. John C. Wister.)
3 plants of *Clinopodium*. (Given by Mr. W. M. Buswell.)
45 plants for Fern Garden. (Given by Mr. H. E. Ransier.)
1 plant for Iris Garden. (Given by Mr. Frank H. Presby.)
2 plants for Iris Garden. (Given by Mrs. W. G. Du Mont.)
4 plants for Iris Garden. (Given by John Lewis Childs Inc.)
16 plants for Iris Garden. (Given by Mrs. Horatio G. Lloyd.)
11 plants for Iris Garden. (Given by Mr. J. Marion Shull.)
2 plants for Iris Garden. (Given by Mr. Geo. N. Smith.)
6 plants for Iris Garden. (Given by Bobbink and Atkins.)
104 plants for Iris Garden. (Given by Mr. John C. Wister.)
16 plants of *Hymenocallis*. (By exchange with Mr. W. Kimball.)
107 plants for Conservatories. (By exchange with U. S. Nat. Museum through Dr. J. N. Rose.)
3 plants of *Zamia*. (By exchange with Mr. Bathusa through Dr. J. K. Small.)
4 plants of *Hymenocallis*. (By exchange with Mr. J. Arthur Harris.)
6 plants for Conservatories. (By exchange with Florida Wild Life League.)
44 plants derived from seed
1 pkt. of seed. (Given by Mr. Samuel F. Clark.)
1 pkt. of *Alpinia* seed. (Collected by Dr. N. L. Britton.)
1 pkt. of seed. (By exchange with Mr. Hubert Buckley.)
1 pkt. of seed. (By exchange with U. S. Dept. Agric.)
70 pkts. of seed. (By exchange with B. G., Oxford, England.)
200 pkts. of seed. (By exchange with B. G., La Mortola, Italy.)

50 Dahlia plants, 24 varieties. (Given by Judge Josiah T. Marean.)
27 Dahlia roots, 18 varieties. (Given by J. J. Broomall.)
20 Dahlia roots, 15 varieties. (By exchange with Mr. F. P. Quinby.)
13 Dahlia roots and plants, 13 varieties. (Given by C. Louis Alling.)
12 Dahlia roots and plants, 7 varieties. (Given by Mr. Alt F. Clark.)
11 Dahlia roots, 6 varieties. (Given by The Dahlia Farm.)
10 Dahlia roots, 9 varieties. (Given by Slocombe Dahlia Gardens.)
10 Dahlia roots, 8 varieties. (By exchange with Dr. M. A. Howe.)
9 Dahlia roots, 7 varieties. (By exchange with Mrs. S. T. Cushing.)
7 Dahlia roots and plants, 7 varieties. (Given by Mr. Alfred E. Doty.)
7 Dahlia roots, 6 varieties. (Given by Mills & Co.)
7 Dahlia plants, 3 varieties. (Given by R. Vincent, Jr. & Sons.)
6 Dahlia roots, 6 varieties. (Given by Mr. Emmett Dove.)
6 Dahlia roots, 4 varieties. (Given by Dahliadel Nurseries.)
6 Dahlia roots and plants, 2 varieties. (Given by Mr. W. J. Matheson.)
5 Dahlia roots, 4 varieties. (Given by Mrs. A. F. Story.)
5 Dahlia roots, 3 varieties. (Given by Capt. J. R. Howell.)
3 Dahlia roots, 3 varieties. (Given by Mrs. Chas. H. Stout.)
3 Dahlia roots, 3 varieties. (By exchange with Miss Margaret S. Brown.)
2 Dahlia roots, 2 varieties. (By exchange with Miss Rosalie Weikert.)
2 Dahlia roots, 1 variety. (Given by George Smith & Sons.)
2 Dahlia roots, 2 varieties. (By exchange with Prof. J. B. S. Norton.)
2 Dahlia roots, 2 varieties. (Given by the Garden Club of Ridgewood, N. J.)
2 Dahlia roots, 2 varieties. (By exchange with Mr. L. H. Du Bois.)
2 Dahlia plants, 2 varieties. (Given by Mr. C. Frey.)
1 Dahlia root. (By exchange with Mr. Otto Pfannkuchen.)
1 Dahlia root. (Given by Dr. W. A. Orton.)
4 plants of *Dahlia imperalis* for Conservatories. (Given by Mr. W. J. Matheson.)
71 plants from Florida. (Collected by Dr. J. K. Small & De Winkeler.)
273 Lily Bulbs. (Purchased.)
3 plants of *Dracaena* for Conservatories. (By exchange with B. E. Blaine Bros.)
16 plants. (By exchange with U. S. Dept. Agric.)
48 plants for Conservatories. (By exchange with U. S. Nat. Museum through Dr. J. N. Rose.)
1 plant of *Opuntia*. (By exchange with Mr. E. T. Wherry.)
4 bulbs of *Oxalis tuberosa*. (By exchange with U. S. Dept. Agric.)
4 bulbs for Conservatories. (By exchange with U. S. Nat. Museum through Dr. J. N. Rose.)
14 plants derived from seeds.
6 plants for Conservatories. (Given by Mr. H. W. Becker.)
4 plants for Conservatories. (Given by Mr. Charles Fett.)
4 plants for Conservatories. (Given by Julius Roehrs Co.)
1 plant of *Pandanus Victoria* for Conservatories. (Given by Mr. Wm. B. Thompson.)

67 plants of Cacti. (By exchange with U. S. Nat. Museum through Dr. J. N. Rose.)

102 plants for Conservatories (By exchange with Mr. Samuel Untermyer.)

15 plants of *Opuntia Drummondii.* (Collected by Mr. D. W. Gross.)

3 plants of *Juglans major* for Nurseries. (Given by Mr. Willard G. Bixby.)

97 plants of *Hicoria* for Arboretum. (Given by Mr. J. F. Jones.)

2 plants of *Hymenocallis* for Conservatories. (Given by Bro. W. Wolf.)

10 plants of *Hicoria* for Arboretum. (Given by Vincennes Nurseries.)

3 plants for Nurseries. (Given by Mr. J. McCarthy.)

1 plant of *Acer rubrum* for Nurseries. (Given by Mr. Chas. C. Dean.)

13 plants of *Cotoneaster* for Fruticetum. (Given by Cottage Gardens Co.)

1 plant of *Iris flavissima* for Iris Garden. (Given by Mrs. C. S. McKinney.)

10 plants of *Hicoria* for Arboretum. (Given by The McCoy Nut Nurseries.)

8 plants of *Hymenocallis* for Conservatories. (Given by Bro. Knapke.)

3 plants for Nurseries. (Given by Mr. E. P. Martin.)

1 plant for Conservatories. (Given by Mrs. Geo. H. Plympton.)

1 plant of *Vaccinium crassifolium.* (Given by Gertrude W. Wilkens.)

15 plants of *Opuntia.* (Collected by Mr. D. W. Gross.)

1 plant of *Pediocactus Simpsonii.* (Collected by Mr. Joseph A. Holmes.)

8400 Gladiolus bulbs. (Given by Mr. A. E. Kunderd.)

6 plants of *Vitis* for Nurseries. (Purchased.)

27 plants for Arboretum and Nurseries. (Purchased.)

49 plants for Conservatories. (By exchange with U. S. Nat. Museum through Dr. J. N. Rose.)

53 plants for Conservatories. (By exchange with Missouri Bot. Garden.)

31 plants for Conservatories. (Given by Mr. P. H. Rolfs.)

16 plants of *Fragaria vesca.* (Given by Mrs. J. W. Martens.)

9 plants for Iris Garden. (Given by Mr. D. M. Andrews.)

153 specimens of Japanese Iris. (Given by Mr. Bertram H. Farr.)

5 plants for *Dionaea muscipula.* (Given by Mr. George Tilles.)

4 plants for Iris Garden. (Given by Mrs. T. Bodley.)

218 plants. (Purchased.)

50 plants for Conservatories. (By exchange with U. S. Nat. Museum through Dr. J. N. Rose.)

1 plant of *Rhododendron.* (By exchange with U. S. Dept. Agric.)

2 plants of *Hymenocallis* for Conservatories. (By exchange with U. S. Dept. Agric.)

52 Florida plants. (Collected by Dr. J. K. Small.)

4 plants of *Opuntia* from Florida. (Collected by J. K. Small and De Winkeler.)

1 packet of *Granadilla* seed. (Given by Mrs. Barsett.)

1 packet of seed. (Given by Mr. F. F. Von Vilmorsky.)

90 packets of seed. (By exchange with B. G., Ottawa, Canada.)

1 packet of seed. (By exchange with Mr. H. G. Wolfgang.)

1 packet of *Linum* Seed. (By exchange with Mr. M. Hawks.)

3 packets of seed. (By exchange with Mr. D. T. A. Cockerell.)

3 packets of seed. (By exchange with Mr. Samuel Untermyer.)

62 packets of seed. (Purchased.)

Provisions for
Benefactors, Patrons, Fellows, Fellowship Members, Sustaining Members, Annual Members and Life Members

---

1. Benefactors

The contribution of $25,000.00 or more to the funds of the Garden by gift or by bequest shall entitle the contributor to be a benefactor of the Garden.

2. Patrons

The contribution of $5000.00 or more to the funds of the Garden by gift or by bequest shall entitle the contributor to be a patron of the Garden.

3. Fellows for Life

The contribution of $1000.00 or more to the funds of the Garden at any one time shall entitle the contributor to be a fellow for life of the Garden.

4. Fellowship Members

Fellowship members pay $100.00 or more annually and become fellows for life when their payments aggregate $1000.00.

5. Sustaining Members

Sustaining members pay from $25.00 to $100.00 annually and become fellows for life when their payments aggregate $1000.00.

6. Annual Members

Annual members pay an annual fee of $10.00.
All members are entitled to the following privileges:

1. Tickets to all lectures given under the auspices of the Board of Managers.
2. Invitations to all exhibitions given under the auspices of the Board of Managers.
3. A copy of all handbooks published by the Garden.
4. A copy of all annual reports and Bulletins.
5. A copy of the monthly Journal.
6. Privileges of the Board Room.

7. Life Members

Annual members may become Life Members by the payment of a fee of $250.00.

Information

Members are invited to ask any questions they desire to have answered on botanical or horticultural subjects. Docents will accompany any members through the grounds and buildings any week day, leaving Museum Building at 3 o'clock.

Form of Bequest

I hereby bequeath to the New York Botanical Garden incorporated under the Laws of New York, Chapter 285 of 1891, the sum of

Vol. XXII   October, 1921   No. 262

JOURNAL

OF

The New York Botanical Garden

EDITOR

R. S. WILLIAMS

Administrative Assistant

CONTENTS

PRICE $1.00 A YEAR; 10 CENTS A COPY

PUBLISHED FOR THE GARDEN
AT 8 WEST KING STREET, LANCASTER, PA
INTELLIGENCER PRINTING COMPANY

JOURNAL

OF

The New York Botanical Garden

| VOL. XXII | October, 1921 | No. 262 |

BOTANICAL COLLECTING IN FRENCH GUIANA.

On the 10th of April this year I sailed from Port of Spain in the French mail steamer, St. Raphael, to undertake a three months' botanical exploration of any part of the Colony which I was to decide upon myself after arrival at Cayenne, the capital.

I arrived at Cayenne on April 15th (five days after I left Trinidad) and landed in rain. I had of course previously received sanction from the then Governor of Trinidad and Tobago, Sir John Chancellor, R. E., K. C. M. G., D. S. O., who had in turn obtained the authority of the Secretary of State for the British Colonies, London, England, to permit of my going, as this was special work outside the scope of the Department of Agriculture of Trinidad and Tobago to which I am attached as Assistant Botanist. And I desire to place on record that it was through the kindly recommendation of Mr. W. G. Freeman, Director, in the first instance, that I was able to obtain four months' leave for the purpose. Also to Mr. Wm. Nowell, who was Acting Director after Mr. Freeman had gone on vacation leave to England, I wish to express gratitude and appreciation for his good offices in the matter.

It was at the instance of Dr. N. L. Britton, Director-in-Chief of the New York Botanical Garden, that I was sent on this important errand to gather botanical specimens for the New York Botanical Garden, the United States National Herbarium, and the Gray Herbarium at Harvard University. When I got to Cayenne, after having had an opportunity to look around, I decided to make the capital my temporary home, as it was noted that plants, of many families, were at hand both in the streets, on its walls, and in the canals which form no inconsiderable proportion of the areas which go to make up the town of Cayenne. The major portion of the land for miles outside of it consists of swamps, and numbers of persons dwell in houses built at the side of, and in some instances, right into the swamps themselves. In these places grow vast numbers of mangroves, especially *Avicennia nitida* and rushes of the genus *Eleocharis*. Hitherto few botanists have collected systematically in French Guiana, notably J. B. Patris in the year 1795, Aublet earlier during the same century and Lagot about 1855. Patris' plants went to Switzerland and Aublet's and Lagot's to Paris. Apart from these few collections little appears to be known of the botany of any part of French Guiana, hence Dr. Britton's anxiety to acquire some of the wild plants which grow in that fertile and interesting country for the three institutions referred to above.

After my arrival at Cayenne and due to complications of the customs regulations, I found that the French steamer St. Raphael by which I had come, had not landed my collecting outfit and baggage, but had over-carried the lot to Martinique. After a delay of a month, in consequence of the miscarriage, these goods and chattels were sent back to Cayenne by a French cargo boat. Here a problem presented itself and no alternative was left me but to seek for old newspapers to use as substitutes meanwhile for the driers that had gone to Martinique instead of being landed at Cayenne on April 15th. I managed with the help of the British Vice-Consul, Monsieur F. Rambaud, Mr. John Grosvenor of the British Consulate, and Monsieur Romney, of the Government Secretary's office, to accumulate a sufficient quantity of this class of paper and thus was able to begin botanical collecting without waiting until the proper papers were returned to Cayenne.

It rained throughout the period I was in French Guiana, which made my stay there run into four months instead of three, as originally intended, it being found impossible to return earlier owing to rains, steamer irregularities, and other unforeseen disadvantages. The drying of specimens in the midst of so much humidity was a difficult operation, especially with soft-tissued plants such as the aroids. Mosquitoes were a terrible drawback, for they attacked me in hundreds and at times in thousands, seeming to come from all points of the compass, including in particular the ground upon which one had to stand or walk, as well as from the shrubs and trees that furnished a portion of the herbarium specimens. How to collect in the midst of such pests with a modicum of comfort, is a question well worth consideration for future action in those immense swamp districts. Perhaps I may be pardoned if I dwell on these enemies a moment so as to emphasize the painful ordeal a collector has to go through whilst engaged in his work in that part of the world. The worst of the mosquitoes is a big, black fellow known as "Mac," which darts at you deliberately and extracts blood immediately from bare arm, hand, head and even through clothing. I know something of mosquitoes, but the French Guiana ones beat anything I ever met in the West Indian Islands of Carriacou, Grenada, Tobago and Trinidad.

The Wet Season of French Guiana comes in at a time when it is the Dry Season of Trinidad and other islands of the Caribbean Sea, so that if one wishes to botanize when the weather is dry, the visitor would have to be there from July or August until December or January.

Due to the fact that I had received directions from Dr. Britton to limit the area for field work and to pick up everything I came across, weed or otherwise, which might be in flower or fruit, Cayenne, as stated above, was made my headquarters. My duty therefore was to overlook nothing that came in my way, either as street weeds or trees of the country, or its mosses, fungi, vines or shrubs; hence I was not restricted to special groups, or species, that were indigenous in the districts where I was occupied in plant-study and plant collecting. I should like to repeat that Cayenne is a town of swamps that extend for miles to the eastward. Long before you enter the harbor you pass millions of mangrove trees with no sign of sandy sea shores. Mud everywhere! Upon landing and after inspecting the trees in and out of town, I observed with surprise the total absence of wild pines (Bromeliaceae), not an individual appearing on a single tree.

True, I met a few clumps of the long-leaved terrestrial species (*Nidularium Karatas*) on Matabon hill, but these were neither flowering nor fruiting: that was the only representative of the family which was seen by me. This is a fact not easily explained, because unusually large aerial masses of orchids of the genus *Epidendrum* (not in flower but possibly the common *E. fragrans* and *E. ciliare*) had become attached to bare trunks of cabbage palms. *Epidendrum stenopetalum*, which was blossoming freely and had bigger flowers than the Trinidad form, and *Epidendrum strobiliferum* were common on mango trees about Cayenne, as also *Polypodium incanum*, a well-known tree-loving fern. I saw one other orchid on trees in sparing quantities but that was not flowering, the leaves were terete and suggested small plants of *Oncidium Cebolleta*, and yet I feel convinced it was not that plant. The orchid flora is exceedingly poor, either in terrestrial or epiphytal species, in the vicinity of Cayenne. Plants of other groups are strong, whether they be on walls or in the ground, large shrubs being a common feature on the sides and tops of masonry structures. Then, why should Bromeliaceae be absent?

A few of the more prominent plants of the town and its immediate environment are *Solanum torvum, Jatropha urens, Lantana Camara, Eleusine indica, Tragia volubilis* (climbing stinging nettle), *Piper marginatum, Leucaena glauca* and *Mimosa pigra*. The swamps of course have their own vegetation, the more conspicuous being sedges and innumerable mangroves. Just outside the limits of Cayenne the land takes on the character of natural savannahs which are unsuited for agricultural purposes. Half a mile along the St. Madeleine Road begin savannah stretches which support such plants as sedges, grasses, *Adiantum, Mabea, Hirtella, Inga, Annona, Psidium, Euphorbia, Canavalia, Maximiliana*, Melastomaceae and Rubiaceae. In the swamps farther out on the same road are mangroves and the tall, massive "bally-hoo" (*Ravenala guianensis*). This is the principal place where the refuse of Cayenne is daily deposited, being brought hither by special carts driven by convicts, causing, in consequence, this part of the country to have an atmosphere of extreme unpleasantness to the collector, although the residents admit they do not notice it, as they are "used to it." It was here, aggravated by mosquito attacks and rains, that much of my collecting had to be done. And it was here that the blossoming, shrubby Melastomaceae were so grand a sight that it made one forget everything else, while admiring their beauty and floriferousness. Beyond in the swamps, quantities of the tall *Thalia geniculata* were in full vigor and blossom, and it was near there where a few plants of the tall grass *Chaetochloa Poiretiana* were seen and nowhere else. The main roads are suited to car driving and by this means distances can be traversed over that portion of French Guiana which forms the island of Cayenne and upon which the capital, bearing the same name is included. "Mango" and Badouel are sandy parts of the country situated in proximity to the town, designated by botanists and naturalists as savannah areas; poor lands unsuited generally for supporting useful plants, either for industrial or food purposes. Large, old mango trees are numerous and the presence of these fruit-trees gave the name of "Mango" to the place in the first instance, I was informed. Many plants of interest are found hereabouts, *Isertia*, grasses, Scrophulariaceae and Apocynaceae, attracting first attention. In the adjacent districts some low hills, two to three hundred feet high, come into prominence and upon these elevated lands grow forest trees which throw dense shade to the ground below. I had an opportunity of botanizing upon three of them, (a) Matabon hill, where a reservoir is to be seen, (b) the hill above Grant's road and (c) another hill close to a rum distillery, parts of each of the three being washed by the open sea. The vegetation is thick and is composed of trees, shrubs, large vines and ferns, representing but a few species, one or two Scitamineae, some grasses and a few sedges.

One terrestrial orchid was collected; it had a solitary fruit, was about two feet tall and this was the only ground orchid I saw throughout my four months' visit. No *Leiphaimos* (*Voyria*) was observed, notwithstanding the ideal conditions found on these forest-clad hills for the growth of such plants. Two things may perhaps be signalled out for special notice just at this point, my number 799 (*Helosis* sp.) which had for its home a place under dense shade at Matabon, and a climbing aroid whose shining green leaves and pink-colored shaggy petioles were exceptionally pretty. On the hill above Grant's Road to which I frequently went and among the arborescent flora grew trees with round green fruits looking not unlike a West African rubber-producing *Landolphia*. The fruits were found to contain a great deal of white sap which was of a tenacious stickiness, and from all parts of the tree, when freshly cut, flowed this colored sap. Herbarium specimens, while drying, had a rather pleasant smell and the cut portions turned dark brown.

In Cayenne there are two public avenues, one formed of full-grown mango trees where boys and men stone the trees unmercifully during the time they are fruiting, and the other running up past the cemetery and turning off towards Badouel and the so-called Botanic Garden. This avenue has been formed of locust (*Hymenaea*) with a sprinkling of *Carapa guianensis*, the "crappo tree" of Trinidad. The small area called the "Botanic Garden" is full of bush and weed; boys and men shoot the few wild birds unhindered, there being no license for firearms and no check by the authorities to the killing of those of utility and beauty. Freedom of action in this respect would seem to exist in and out of Cayenne.

French Guiana is a magnificent fruit country. Its bananas, oranges, mangoes and plantains are of the best, not only in size but in quality as well. The lime trees (*Citrus*) are healthy and free of wither-tip and other pernicious diseases. Small sugar-cane fields were looking in healthy condition, but agricultural matters in that part of the Guianas are backward. So far as could be ascertained, the cultivating of canes is primarily for the purpose of producing the favorite drink of the place, known as taffia there and rum elsewhere. The three chief industries of French Guiana, apart from rum-making, are gold-digging first, followed by Balata-tree bleeding for gum and the making of a liquid essence from a timber tree known as "bois rose." At the time of my visit these industries were not thriving, many people being out of work as the result of employers closing down their respective business places, yet that made no difference in the high cost of food and clothing at Cayenne.

The principal water supply of the town comes from a considerable distance, where the wild vegetation is fairly luxuriant, but I had only one opportunity of seeing it, and that was the day before I left Cayenne on August 1st; it was there I saw a large Cannon-ball tree (*Couroupita*). In one of the swamps there was growing and flowering a tall grass with green infloresence which hung from the tips of the culms. Unfortunately the water was too deep and the time too limited to allow of my collecting specimens. A *Montrichardia* was also seen which matched well with the Trinidad *M. aculeata*. Scarcely a butterfly or other large insect was seen either on this or any other day and birds appeared scarce.

During the time I was at Cayenne I saw neither moth nor allied insect attracted to the electric lights at night. These blanks in the fauna puzzled me as much as did the absence of the tree-loving Bromeliaceae. Mud, mire and unpleasant odors usually confronted one, so that it was refreshing to see a piece of sandy sea shore as that which skirts the end of Grant's road, Matabon. It was there that "drift seeds" were picked up among swarms of biting sand flies; off the beach, islets stand in view and upon one of the smaller ones upright growing cactuses were seen.

In closing this narrative I must not omit to refer to the kindness rendered me by His Excellency, the Governor of French Guiana; Monsieur Romney of the Chief Government Secretary's Office; Monsieur Magney, Agent of the French Line of Steamers; Monsieur F. Rambaud, the British Vice-Consul; and Mr. John Grosvenor of the British Consulate. I also received courtesies and civilities from every government official and private resident I met.

The two chief fodder grasses cultivated at Cayenne are Pará grass and rice grass (*Chaetochloa* sp.) Guiana grass seems to be unknown. Palms furnish large lots of fruits that are bought up readily in the market as food items in their season.

The town of Cayenne has wide streets, good residences, a splendid water supply for drinking purposes and a daily ice out-put. At night electric lights are seen throughout the place. Due to easterly winds the air is cool and pleasant even at midday in the shade.

In their entirety my collections were shipped to Dr. Britton from Port of Spain to New York. One other item ought to be added and that is the rare occurrence of a two-trunk cabbage-palm some sixty feet in height among the rest of the Oreodoxas outside the Hospital of St. Paul.

W. E. BROADWAY

HOW TO GROW RHODODENDRONS.[1]

Rhododendrons can be grown much more easily in the greater part of Southeastern New York, than in any part of Western New York, as lime is present in the soil throughout the most of Western New York. By Western New York I refer more particularly to the region around Rochester, Batavia and Buffalo. In most instances where success has been obtained in Western New York in the growing of Rhododendrons the soil has been excavated to a depth of several feet and replaced with humus. This of course is an expensive operation.

During the past thirty years collections of Rhododendrons were more or less severely injured throughout the greater part of the Northeastern States by the severely cold winters of 1903 and 1904, 1917 and 1918, and 1919 and 1920, causing a vast amount of injury in the burning of the foliage, half-killing many, and killing large numbers out-right. Many growers were much discouraged because a number of those forms that were considered hardy were in many instances severely injured or killed. Azalea species, and numerous hybrids, suffered comparatively very little, and presented a good display of flowers, and only a limited number of their flower buds were blasted with the intense cold.

The so-called Catawbiense hybrid Rhododendrons, in a somewhat limited number, which are familiar in many gardens, and a few American, Caucasian and Asiatic species are the only forms that have been at all successful in cultivation in the Northeastern States, and we do not think their cultivation should be abandoned on account of a few "set backs". The origin of these Catawbiense hybrids is more or less obscure and involved, but whatever hardihood they have is surely due to the presence of the blood of our native *R. catawbiense*. Very few named forms have been added to these hardy Catawbiense hybrids recently.

A number of hybrids of reputed hardihood have resulted from the introduction of the blood of *R. caucasicum* with other forms. Boule de Neige is the only one of these we have had experience with. It is a compact plant with white flowers, very hardy, and a beautiful form.

[1] Public lecture at the New York Botanical Garden, September 3, 1921.

The Caucasian *R. Smirnowii* has been crossed with *R. catawbiense* at the Arnold Arboretum, and the progeny is said to be quite hardy, The flowers vary from clear pink to deep rose. In this there is excellent promise of a race of important garden plants. *R. Smirnowii*, in its typical form, is a handsome flowering plant but it needs considerable shade.

R. carolinianum is a very decorative garden plant, of small compact habit, with flowers varying from pale rosy-purple to almost white, and has a good, hardy constitution. It is a most desirable plant for gardens and parks, and fortunately is now being offered in large quantities by some American nurserymen.

R. ferrugineum, *R. hirsutum* and *R. myrtifolium* on account of their low compact habits are excellent garden plants. Their flowers are pretty, but not conspicuous. It is true that with us their leaves are occasionally somewhat scorched by sun and wind in late winter, but they soon recover with their new growths.

Rhododendron mucronulatum is about the earliest flowering form with us. It has an upright-growing habit. The flowers are rose-colored with a tinge of lilac, and it blooms from the end of March to April 1st. The blossoms are sometimes injured by severe spring freezing.

It is to be regretted that the large number of new species discovered by Henry E. Wilson in Western China will not thrive in any part of the Northeastern States. They do well on some parts of the Pacific Coast and in Great Britain and Ireland.

Some botanists now place what we all know as Azaleas, with the deciduous leaves, generically under *Rhododendron*. Azaleas appear externally to be so different from Rhododendrons that it is difficult for gardeners to become accustomed to calling them Rhododendrons. In an extraordinary wealth of varied rich colors, the hardy species and hybrid Azaleas cannot be excelled by any shrubby garden plants. The late George H. Ellwanger said of Azaleas in the "Garden's Story," "The soft tints of buff, sulphur and primrose, the dazzling shades of apricot, salmon, orange and vermilion, are always a fresh revelation of color. They have no parallel amongst flowers, and exist only in opals, sunset skies and the flush of autumn woods." This may appear exaggerated, but when a large collection of the so-called Ghent hybrids, and different species of Azaleas are assembled together in a ravine, and well established plants are in full bloom, Mr. Ellwanger's description is not overdrawn.

The American Azaleas are lovely flowering plants, and should be extensively planted. Seventeen species are now credited to North America. Eight of these are more or less in cultivation. Perhaps one of the most beautiful is *Rhododendron calendulaceum* distributed freely on the Appalachian range. The flowers vary from light yellow to deep orange-scarlet ánd the plant flowers freely when two and one half to three feet in height.

The status of a beautiful native Azalea has recently been brought to light, and it is now known as *Rhododendron roseum*. This plant has formerly been known as *Azalea nudiflora*. It is native throughout the greater part of New York State and far beyond. It is a common plant in the lime-stone regions of Western New York. In fact I have found it growing on Onondago limestone rocks with only a few inches of soil and the roots literally adhering to the disintegrated limestone. The flowers vary from light pink to deep rosy pink, and are deliciously fragrant. We have in this Azalea a garden plant of much beauty and hardihood, and adapted to a wide range of soil conditions. It should be raised and distributed in tens of thousands.

R. nudiflorum is now known to have a much more southern distribution than *R. roseum*, and is a rare plant in Western New York, but seems to be quite well distributed in the eastern part of the State. The showy pink flowers are very attractive, but have very little fragrance.

The other species of American Azaleas are important in the order following: *R. Vaseyi*, *R. arborescens*, *R. occidentale*, *R. canadense* and *R. viscosum*.

Of the thirty-four species of Azaleas credited to the Old World in Wilson & Rehder's monograph of Azaleas perhaps not more than ten or twelve species are hardy, outside, as far north as New York State. One of the most hardy and beautiful is *R. japonicum*. The flowers range in color from orange-red to almost red. This was for a long time confused with the more tender *R. molle*. The latter has handsome yellow flowers, but has not the enduring qualities of *R. japonicum*. The handsome hybrid race known as *R. Kosterianum*, in numerous forms, is the result of crossing *R. japonicum* with *R. molle*.

Rhododendron Schlippenbachii promises to become an important garden plant. It is said to be a common shrub in Korea and particularly abundant on the Diamond Mountains. It has a very distinct appearance and can easily be recognized from any other Azalea. It flowered with us last spring for the first time. The large blossoms are clear pink and are quite showy.

Kaempfer's Azalea, now known botanically as *Rhododendron obtusum* var. *Kaempferi*, is a most desirable and beautiful garden plant. Introduced from Japan about thirty years since, it is now fairly well distributed. The flowers usually vary from light red to deep red. It seems to enjoy partial shade, and in the Arnold Arboretum does splendidly under the partial shade of hemlocks, showing a much richer coloring than where it is exposed to the full sun.

The dainty and lovely Kurume Azaleas recently introduced to American gardens from Japan, which reveal a wide range of exquisite shades, are now attracting a great deal of attention and may, perhaps, prove to be hardy in Long Island and southward, outside, as well as in sheltered places around New York. I do not think they would be at all satisfactory in Western New York out of doors.

Amongst the various groups of hardy hybrid Azaleas the so-called Ghent Azaleas are the most desirable. The flowers are, perhaps, not as large as in some of the other hybrid groups, but the tinting is exquiste, and most of them are fragrant.

Somewhere about 1825 a baker at Ghent hybridized some of the American Azaleas amongst themselves, and also with the Pontic Azalea, now known as *Rhododendron luteum*. About the same time similar work was done in England; in all, many hundreds of named forms were produced. Later the Japanese and Chinese Azaleas were hybridized amongst themselves, and with other hybrid groups, in which also the beautiful Pacific coast *Azalea occidentalis* was involved until "the parentage of most of these forms is so mixed that it is impossible to recognise their origin with certainity."

The forms of the Ghent hybrids are perhaps more enduring than the individuals of the other hybrid groups, and some of the best are Gloria Mundi, Altaclarensis, Unique, Cardinal, Daviesei, Viscocephala, Pallas, Fritz Quihoui, General Trauff, Grandeur

Triumphante, Ignea Nova, Joseph N. Bauman, Julda Schipp, Madame Moser and Pucella.

As already intimated the soil in Highland Park, Rochester, contains lime. The subsoil is light, sandy loam, on gravel, and well drained naturally. This soil was excavated to a depth of two and a half to three feet and replaced with humus which was found conviently near in a "Kettle-hole," with which cow dung was liberally mixed. All attempts to grow Azaleas previously, without this preparation, were absolute failures. The plants are well protected from the sweep of the north-west and north winds by sloping banks, and additional planting of evergreens on the east and west sides, adds to the protection. No over-head covering is attempted in winter. Late in the autumn the entire area is heavily mulched with leaves to a depth of eight or nine inches, which are allowed to remain throughout the year, and this is repeated annually. Extreme watchfulness must be exercised in preventing the humus from becoming dry, because if this is permitted, it is extremely difficult to restore it to the point of saturation, and the plants will suffer and pass through the winter badly.

I might say here that in Durand-Eastman Park, ten miles north of the city of Rochester, on Lake Ontario, and north of what is known as the "Ridge Road" the soil contains only a faint trace of lime, and we have found to our great satisfaction that the American Azaleas and Rhododendrons do remarkably well in the light sandy soil, without more than a surface mulching of leaves or decayed manure. It is only during the past few years this fact has been known.

In regions where Rhododendrons and Azaleas will thrive, where no lime is present in the soil, it is an entirely unnecessary expense to excavate and replace with soil of a peaty nature, as I know is sometimes done. If the soil is a light sandy loam and well drained, all that is necessary is to trench it to a depth of two and a half to three feet, and incorporate a liberal amount of humus and cow dung in the surface and Rhododendrons will be happy. Protection from the sweep of the cold northwest and north winds, which are so sear and penetrating in February and first of March, is essential by the lay of the land or by some other means. If tall trees are adjacent to break a little of the winter sunshine it may help them considerably. I believe however if they are thoroughly mulched, and the roots are in the right status of moisture, winter sunshine has but little effect on them.

The propagation of Rhododendrons and Azaleas is now a very important matter in this country since the government prohibited the importation of any plants with balls of earth.

Rhododendrons are not difficult to raise from seeds and our experience in raising seedlings from the best colored Catawbiense hybrids has been most gratifying. We have raised a number of forms that in our opinion are about as good as the named forms. Very fine pink and lavender-pink shades have been obtained. We have one clear pink form that flowers ahead of any of the named Catawbiense hybrids. We strongly urge the raising of seedlings from these Catawbiense hybrids, particularly where a large collection is assembled, as no doubt cross pollination is effected naturally. They will flower in from five to six years from seed.

It is hoped that American nuserymen wherever they have the opportunity, will propogate Rhododendrons and Azaleas extensively, and I do not see any reason why the typical *R. catawbiense* cannot be used as a stock for Rhododendrons instead of *R. ponticum*. We have used *R. catawbiense* in a limited quantity and so far it appears to be satisfactory. *R. ponticum* is the common stock for grafting Rhododendrons on the continent of Europe. It is known to be tender and we believe for cold regions that *R. catawbiens'* would be better suited. Stocks should be potted in spring. By autumn they will be well rooted. Move them from a cold frame into the greenhouse in mid-winter, and after they are well rooted, the cions can be united towards the base by side or veneer grafting, and tied with strong cotton. The cutting back of the stock should be done very gradually after the union has been thoroughly established. Some propagators do not head the stocks back entirely until the following year.

Seeds of either Rhododendrons or Azaleas should be sown in the greenhouse as soon as ripe, in small flats or pans, well drained, on a finely prepared surface of sandy, peaty soil. They should be kept close to the glass and shaded from direct sunshine until they are well germinated. Care must be exercised not to over-water or they will damp readily, particularly in dull winter weather. By spring they will be large enough to transplant to other well drained flats or pans. It is a good plan to keep these pans or flats in lightly shaded frames throughout the summer. In late summer or early autumn they will be ready to shift into two or two and one half inch pots and carried throughout the winter in a cool greenhouse temperature. The following spring they can be transplanted from the pots into nursery beds.

JOHN DUNBAR,
Rochester, New York.

NOTES, NEWS AND COMMENT

The flowering of the dahlia border was brought to an abrupt close by killing frosts on the mornings of October 26 and 27. The advent of the first killing frost wa: the earliest in the four years of the border's existence. In 1918 it was killed on November 7; in 1919, on November 10; and in 1920, on November 13. The faint hope that an annual increment of three days in the length of the season was being established was doomed to disappointment. On account of the light rainfall of the summer and autumn, the plants were less luxuriant and the flowers less profuse than in the more copiously watered seasons of 1919 and 1920. The exhibit, however, included many new and superior varieties and the border, as in the three previous years, attracted much attention and did much to keep the people of New York and vicinity in touch with the latest perfections in dahlia-breeding. The beautiful novelties contributed by Judge Josiah T. Marean of Green's Farms, Conn., and by J. J. Broomall of Eagle Rock, California, were especially noteworthy.

Miss Sophie Satin, a mycologist and former member of the University of Moscow, but at present residing in New York, has recently paid a number of visits to the Botanical Garden Museum for the purpose of looking over our mycological collections.

Meteorology for October:—The total precipitation for the month was 0.58 inch. The maximum temperatures recorded for each week were as follows: 75° on the 7th, *78½° on the 11th*, 76° on the 19th and 66° on the 28th. The minimum temperatures were: 37° on the 9th, 36° on the 13th, 35° on the 24th and *28° on the 27th*. The first killing frost of the autumn was on the morning of the 26th, when a temperature of 29½° was recorded.

ACCESSIONS

MUSEUM AND HERBARIUM

1 specimen of *Cepaluros virescens* from Mississippi. (Given by Professor L. E. Mills.)

50 specimens of marine algae from Bermuda. (Given by Dr. A. B. Hervey.)

26 specimens of flowering plants from the Bolivian Andes. (By exchange with Harvard University.)

1279 specimens of flowering plants from Haiti. (By exchange with the United States National Museum.)

105 specimens of flowering plants from Utah. (By exchange with Professor A. O. Garrett).

121 specimens of flowering plants from Arizona. (Distributed by Mr. W. N. Clute.)

5 specimens of flowering plants from southern Utah. (Given by Dr. Harry Hapeman.)

1 specimen of *Osmunda claytoniana* from Vermont. (Given by Mrs. W. E. Mack.)

48 photographs of plants, mostly cacti. (By exchange with the United States National Museum.)

1 lantern slide. (By exchange with the Brooklyn Botanic Garden.)

11 specimens of flowering plants. (By exchange with the Royal Botanic Gardens, Kew, England.)

49 specimens of flowering plants from the Peruvian Andes. (By exchange with Harvard University.)

10 specimens of hepaticae from New England. (Given by Miss Annie Lorenz.)

3 specimens of orchids from eastern North America. (Given by Dr. H. M. Denslow.)

36 specimens of flowering plants from South America. (By exchange with Harvard University.)

30 photographs of cacti. (By exchange with the United States National Museum.)

53 specimens of mosses from Trinidad. (Collected by Mrs. N. L. Britton.)

56 specimens of mosses from Hawaii. (By exchange with Rev. David Lillie.)

178 specimens of mosses from Georgia, Alabama, Florida, Missouri, Colo-

rado, Utah, Alaska, Guatemala, Panama, Venezuela, and Africa. (By exchange with the United States National Museum.)

678 specimens of mosses from Michigan. (By exchange with Dr. George E. Nichols.)

32 specimens of mosses from California and Montana. (By exchange with Miss Alice Eastwood.)

12 specimens of mosses from British Guiana, Maine, and Vermont. (By exchange with Professor Edward B. Chamberlain.)

4 specimens of mosses from Portugal and Madeira. (By exchange with Professor T. D. A. Cockerell.)

1 specimen of moss from California. (By exchange with Miss Dorothy Coker.)

37 specimens of mosses from Java. (By exchange with Dr. A. W. Evans.)

60 specimens of mosses from various localities. (By exchange with Professor J. M. Holzinger.)

865 specimens of flowering plants from North America. (By exchange with the United States National Museum.)

4 specimens of flowering plants from Alabama. (Given by Dr. R. M. Harper.)

1791 specimens of Brazilian plants. (By exchange with the British Museum.)

1 specimen of *Poria* from New York. (By exchange with Dr. H. D. House.)

1 specimen of *Stropharia ambiguum* from California. (By exchange with Alice Eastwood.)

25 specimens of fungi from Porto Rico. (By exchange with Professor F. S. Earle.)

3 specimens of fungi from Albertia, Canada. (By exchange with Mr. N. B. Sanson.)

12 specimens of fungi from Bermuda. (By exchange with Dr. H. H. Whetzel.)

3 polypores from Brazil. (By exchange with Mr. H. M. Curran.)

6000 specimens of flowering and flowerless plants from Trinidad. (Collected by Dr. and Mrs. N. L. Britton.)

625 specimens of ferns and fern-allies from Jamaica. (Collected by Mr. W. R. Maxon.)

164 specimens of flowering plants from Guatemala. (By exchange with Professor W. W. Rowlee.)

250 specimens of flowering plants from Jamaica. (By exchange with the Royal Botanic Gardens, Kew, England.)

23 specimens and photographs of cacti (By exchange with the United States National Museum.)

1 specimen of *Sonchus uliginosus* from Pennsylvania. (Given by Mr. E. A. Rau.)

18 specimens of mosses from Santo Domingo. (By exchange with the United States National Museum.)

478 specimens of flowering plants from Martha's Vineyard. (By exchange with Harvard University.)

The New York Botanical Garden

Journal of the New York Botanical Garden, monthly, illustrated, containing notes, news, and non-technical articles of general interest. Free to all members of the Garden. To others, 10 cents a copy; $1.00 a year. [Not offered in exchange.] Now in its twenty-second volume.

Mycologia, bimonthly, illustrated in color and otherwise; devoted to fungi, including lichens; containing technical articles and news and notes of general interest, and an index to current American mycological literature. $4.00 a year; single copies not for sale. [Not offered in exchange.] Now in its thirteenth volume.

Addisonia, quarterly, devoted exclusively to colored plates accompanied by popular descriptions of flowering plants; eight plates in each number, thirty-two in each volume. Subscription price, $10.00 a year. [Not offered in exchange.] Now in its sixth volume.

Bulletin of the New York Botanical Garden, containing the annual reports of the Director-in-Chief and other official documents, and technical articles embodying results of investigations carried out in the Garden. Free to all members of the Garden; to others, $3.00 per volume. Now in its tenth volume.

North American Flora. Descriptions of the wild plants of North America, including Greenland, the West Indies, and Central America. Planned to be completed in 34 volumes. Roy. 8vo. Each volume to consist of four or more parts. Subscription price, $1.50 per part; a limited number of separate parts will be sold for $2.00 each. [Not offered in exchange.]

Vol. 3, part 1, 1910. Nectriaceae—Fimetariaceae.

Vol. 7, part 1, 1906; part 2, 1907; part 3, 1912; part 4, 1920; part 5, 1920; part 6, 1921. Ustilaginaceae—Accidiaceae (pars). (Parts 1 and 2 no longer sold separately.)

Vol. 9 (now complete), parts 1-7, 1907-1916. Polyporaceae—Agaricaceae (pars). (Parts 1-3 no longer sold separately.)

Vol. 10, part 1, 1914; parts 2 and 3, 1917. Agaricaceae (pars).

Vol. 15, parts 1 and 2, 1913. Sphagnaceae—Leucobryaceae.

Vol. 16, part 1, 1909. Ophioglossaceae—Cyatheaceae (pars).

Vol. 17, part 1, 1909; part 2, 1912; part 3, 1915. Typhaceae—Poaceae (pars).

Vol. 21, part 1, 1916; part 2, 1917; part 3, 1918. Chenopodiaceae—Allioniaceae.

Vol. 22, parts 1 and 2, 1905; parts 3 and 4, 1908; part 5, 1913; part 6, 1918. Podostemonaceae—Rosaceae.

Vol. 24, part 1, 1919; part 2, 1920. Fabaceae (pars.)

Vol. 25, part 1, 1907; part 2, 1920; part 3, 1911. Geraniaceae—Burseraceae.

Vol. 29, part 1, 1914. Clethraceae—Ericaceae.

Vol. 32, part 1, 1918. Rubiaceae (pars).

Vol. 34, part 1, 1914; part 2, 1915; part 3, 1916. Carduaceae—Anthemideae.

Memoirs of the New York Botanical Garden. Price to members of the Garden, $1.50 per volume. To others, $3.00. [Not offered in exchange.]

Vol. I. An Annotated Catalogue of the Flora of Montana and the Yellowstone Park, by Per Axel Rydberg. ix + 492 pp., with detailed map. 1900.

Vol. II. The Influence of Light and Darkness upon Growth and Development, by D. T. MacDougal. xvi + 320 pp., with 176 figures. 1903.

Vol. III. Studies of Cretaceous Coniferous Remains from Kreisch New York, by A. Hollick and E. C. Jeffrey. viii + 138 pp., with 29 plates. 1909.

Vol. IV. Effects of the Rays of Radium on Plants, by Charles Stuart Gager. viii + 278 pp., with 73 figures and 14 plates. 1908.

Vol. V. Flora of the Vicinity of New York: A Contribution to Geography, by Norman Taylor. vi + 683 pp., with 9 plates. 1915.

Vol. VI. Papers presented at the Celebration of the Twentieth Anniversary of the New York Botanical Garden. viii + 592 pp., with 43 plates and many text figures. 1916.

Contributions from the New York Botanical Garden. A series of technical papers written by students or members of the staff, and reprinted from journals other than the above. Price, 25 cents each. $5.00 per volume. In the tenth volume.

Vol. XXII   November and December, 1921   No. 263, 264

JOURNAL

OF

The New York Botanical Garden

EDITOR

R. S. WILLIAMS

Administrative Assistant

CONTENTS

PRICE $1.00 A YEAR; 10 CENTS A COPY

PUBLISHED FOR THE GARDEN
AT 8 WEST KING STREET, LANCASTER, PA
INTELLIGENCER PRINTING COMPANY

PLATE 263

In the Deering hammock, Cutler. Large tables of oolitic limestone, right and left, in foreground; pond-apple (*Annona*) trees, hosts of numerous air-plants—orchids and wild-pines, in background. The rim at the top of the tables indicates the maximum normal water-table. Originally a uniform layer of limestone of varying consistency was present. The softer parts were eroded mainly by the fluctuating seasonal water-table; the harder parts resisted erosion, hence the cañons and grottoes on the one hand and the tables on the other. The tables support trees just as large as those on the floor of the hammock.

JOURNAL

OF

The New York Botanical Garden

Vol. XXII   November and December, 1921   No. 263, 264

HISTORIC TRAILS, BY LAND AND BY WATER.

A Record of Exploration in Florida in December 1919.

With Plates 263–266

The prospect of passing, within the space of a few hours, from the dead of winter into the life of summer is naturally fascinating. With this outlook in mind Dr. and Mrs. Britton, Mrs. Small, and the writer left New York en route for Florida about the first of December, 1919.

Repeated killing frosts and freezing weather about New York had ravaged all growing things, except the evergreens, at least in exposed places.

The striking objects in the landscape in emerging from the Hudson River tunnel into New Jersey were the dead stalks and leaves of cat-tails (*Typha*) and those of the common-reed (*Phragmites*). Both of these plants, even the same species, occur as far south on our eastern seaboard as the Everglade Keys and Florida Keys, where they are not only evergreen, but may often be found in flower or in fruit nearly throughout the year.

North of the Mason and Dixon line the hardwood trees were leafless, except in the swamps of New Jersey where the young pin-oaks (*Quercus palustris*), not the old ones, still held their dead, dried leaves. The characteristic bark, clothing the trunk of the white-birch (*Betula populifolia*), was a prominent feature in the landscape of the rocky slopes below New Brunswick, while the smooth-alder (*Alnus serrulata*), in the low places, showed its nascent aments ready to flower with the first real

193

warm spell of the winter. The only green that appeared on the ground was that of the foliage of the winter-annual weeds growing in hedges and in other sheltered places.

In Virginia and North Carolina some green foliage appeared on two or three cultivated shrubs, the Japanese-honeysuckle (*Nintooa japonica*)—often an escape—and privets (*Ligustrum ovalifolium* and *L. amurense*)—usually in hedges—and on one native tree, the sweet-bay (*Magnolia virginiana*), which was especially conspicuous among the other trees on account of its persistent leaves. In addition, a parasitic woody plant, the mistletoe (*Phoradendron flavescens*), furnished greeneries varying from the size of one's head to that of a bushel-basket. It was most abundant on oaks, hickories, and gums.

On barren hills the scrub oaks, and in other places, larger oaks, still held their dead and dried leaves, all of which stood out in strong contrast against scattered groves of the evergreen red-cedar (*Sabina virginiana*).

However, from Virginia to Georgia leaves other than dead ones were in evidence in low places and in swamps; but they were not green. The tulip-tree (*Liriodendron*), the maple (*Acer*), the sweet-gum (*Liquidambar*), and the sour-gum (*Nyssa*) were all brilliant with various shades of yellow and red foliage. Some deciduous-leaved oaks also showed well-colored leaves. Further south, however, one kind of oak was green. It was the live-oak (*Quercus virginiana*) and looked just the same as the live-oak trees in southern Florida. Latitude has little or no effect on the appearance of this tree. Growing with it, occasionally, was the red-mulberry (*Morus rubra*) which was just about devoid of leaves. This tree represents another case in which latitude does not seem to affect the habit of the tree to any appreciable extent. However, instead of being evergreen like the live oak, the mulberry drops its leaves in fall or winter throughout its range, whether it be in Canada or in southern Florida

In southern Georgia the trees in the hammocks of the swamps and along the streams had not put on their fall coloring, at least not to the same extent as those further north.

The extensive marshes, naturally, appeared approximately as they do the year around. Two plants, however, were particularly conspicuous on the flats about the winding channels, strongly contrasted by different shades of color—the one a dark-green rush (*Juncus*), the other a light-green grass (*Spartina*). Moreover, evidences of the cultivated crops of the Georgia lowlands were not wanting, for numerous bright green plants of asparagus were scattered along the railroad, and the dull-yellow stubbles of the rice fields often attracted the eye.

Ditches along the way, filled with a shimmering gold, disclosed the only conspicuous flowering plant then in full bloom. It was a species of bur-marigold (*Bidens*). Blooming as it did in what is to be considered about mid-winter in that latitude, it is hard to decide whether the plant was a belated fall-bloomer or an early spring-bloomer. Associated with the bur-marigold was a very tall, woolly beard-grass (*Erianthus*) with stems six to twelve feet high, terminating in long, silvery, fruiting plumes which, of course, stood far above all the surrounding herbaceous vegetation.

Along the Saint Johns River at Jacksonville more plants were in flower. In a small area there were observed—beard-grass (*Andropogon glomeratus*), water-hyacinth (*Piaropus crassipes*), Indian-shot (*Canna flaccida*), bladder-pod (*Glottidium vesicarium*), aster (*Aster lateriflorus*), groundsel-bush (*Baccharis halimifolia*), and beggar's-ticks (*Bidens leucantha*). There, too, a cultivated tree, the common pear (*Pyrus communis*) and the native toothache-tree or prickly-ash (*Zanthoxylum Clava-Herculis*) were in bloom.

While in the field our collecting headquarters were maintained at the laboratory building of the Plant Introduction Garden of the United States Department of Agriculture, through the courtesy of Dr. David Fairchild. Our field work was made possible by the generous cooperation of Mr. Charles Deering. Mr. Deering's cactus plantation at Buena Vista, moreover, still served as the basis for our studies in the cacti of the eastern United States.

Florida appears to be peculiar among the States in one respect, at least—towns in all parts of that Commonwealth claiming to have "the best location and the best climate."

Notwithstanding this advertised uniformity in place and climate, there is really great diversity in the climate and consequently in the vegetation. The latter, of course, was the magnet that attracted us to the field. Palms and wild-pepper plants were our chief object of search. It was also planned to visit hitherto unexplored points in the southern end of the peninsula.

Mention of a few makers of the early Florida trails may be of interest—among others, prehistoric aborigines; Seminole Indians; adventurers, as de Soto and de Leon; buccaneers, as Black Caesar and Gilbert; shipwrecked refugees, as Dickenson and associates; naturalists, as the Bartrams and Baldwin; land improvement agents, as Turnbull and Perrine; Indian hunters, as Canova and Taylor. Their activities have added greatly to the fascination of subsequent exploration.

Four different historic trails served us in accomplishing the major portions of our field work:

First, the old trail along the eastern coast, from Saint Augustine southward. Parts of this, or what they developed into, date back several centuries. Of late this trail has been transformed into the southern part of the Dixie Highway.

Second, the trail made by the surveyors of past and present generations, in opening up the territory between Miami and Cape Sable. Within the past few years this has been transformed into the Ingraham Highway.

Third, the Fort Bassenger trail, which dates from the Seminole War times. The part in Hungry Land is being improved, that in the Allapattah Flats is in about the same condition it was a century ago. It connected the eastern coast from the vicinity of Jupiter with the Okeechobee region and lower Kissimmee River region, terminating at Fort Bassenger, which was situated on the Kissimmee River about fifteen miles, in a straight line from the northern end of Lake Okeechobee.

Fourth, the so-called inside course on the Florida Reef between points on the lower eastern coast and Key West. The use of this course dates back many years. A history of the craft and personage that traversed it for generations before the power-boat was invented would make interesting reading. Like the trails referred to above, it has been improved in recent years, chiefly by dredging and staking channels.

We left the train at Daytona and proceeded down the eastern coast, by motor-car, to Miami.

Daytona is situated in the remains of a once great hammock. The two most prominent trees still standing in the streets and

Plate 264

Floor of hammock on ancient sand-dune bordering Saint Lucie Sound. Tropical trees comprise this hammock. On the hammock floor grows a typically West Indian wild-pepper (*Peperomia humilis*), shown above. It is sometimes nearly as luxuriant here as it is in the Cape Sable region. The fruits are very viscid, and they were probably brought northward on the feet or plumage of migratory birds. The insert shows a species of wild-pepper (*P. cumulicola*) endemic in Florida. Its distribution is confined to the shell-heaps and habitations of the aborigines.

lots of the town are the live-oak (*Quercus virginiana*) and the cabbage-tree (*Sabal Palmetto*). Among the more abundant epiphytes which grow particularly on the oak are the Spanish moss (*Dendropogon usneoides*) and the little orchid, *Auliza conopsea*. Curiously enough, these two plants, the former with quite simple flowers, the latter with complex flowers, have sepals and petals of the same peculiar shade of green. Immediately south of Daytona, shell-middens, the refuse heaps left by the aborigines, became evident and on these, two additional kinds of trees attract the eye. They are the hickory (*Hicoria floridana*) and a red-cedar which has been included in *Sabina barbadensis*, but which we now feel justified in regarding as different. The insular species bears depressed cones, while the Florida plant has ovoid cones which consequently, are rather longer than thick.

About ten miles south of Daytona we came upon one of the objects of our excursion, namely, a small wild-pepper plant—so far as known at that time the only endemic wild-pepper plant of the continental United States.[1]

This plant is a winter bloomer and was at the height of its flowering season. It was the most conspicuous herbaceous plant on the floor of the hammock, covering the shell-middens for acres. How much more widely it is scattered in that part of the State remains to be learned. This species, like its closest relative in Florida, is now known in only three localities—in all cases on these prehistoric middens: at the locality just mentioned, on shell-mounds at the mouth of the Saint Johns River, where the species was discovered about a century ago, and on apparent refuse heaps, in the mouths of caves, on the western side of Lake Tsala Apopka on the opposite side of the peninsula. In other words, it has been found at two localities near the coast and at one in the interior.

[1] **Peperomia cumulicola** Small. *Piper leptostachyon* Nuttall, Am. Journ. Sci. **5**: 287. 1822. *Peperomia leptosachya* Chapm. Fl. S. U. S. Suppl. 645. 1884. Not Hook. & Arn. 1832.

Plants terrestrial, pubescent with short recurved hairs: leaves bright-green; blades of those of the flowering stems and branches obovate, oval, or suborbicular, 3-veined: spikes about 1 mm. in diameter, mostly 2-8 cm. long: bracts suborbicular, about 0.5 mm. in diameter, erose-edged: berries obovoid, borne in depressions in the spike-rachis.

Shell-mounds and entrances to caverns, eastern Florida and northern part of peninsula. It flowers profusely in December.

Here, south of Daytona, often growing with the *Peperomia*, was our smallest spiderwort, *Tradescantella floridana*. This plant, like the *Peperomia*, was at the height of its flowering season. It grew very vigorously and often formed mats, sometimes with a tangled mass of stems and leaves, an inch in depth. · Our other spiderworts bear conspicuous flowers usually with highly colored corollas. This one, however, bears relatively inconspicuous flowers much less than a quarter of an inch in diameter, with pale, usually white, petals. The corolla always lies horizontally or flat on the masses of green foliage. The calyx is usually dark greenish-purple. Living specimens collected there were brought to the Garden and they have since served as a ground-cover in parts of Conservatory Range number two.

On an elevation on one of the shell-mounds we were confronted with a small grove of the Indian-cherry (*Rhamnus caroliniana*). This was quite a surprise, as the species has its center of distribution in the mountains and adjacent territory and northern Florida was the supposed southern geographic limit of the species. The trees were in fruit. This fruit consists of black, cherry-like berries about as large as a good-sized pea.

We spent the forenoon in the vicinity of the shell-mounds and then proceeding southward and along the coast without much delay, we ended the day on Merritt's Island. All we could accomplish there was a running view of the flora of the fifteen or twenty miles of the southern end of the island. The experience, however, decided us to place it in our list for further exploration at the earliest possible opportunity, for several dozen shrubs and trees typical of tropical regions were found in the various hammocks along the trail.

One of the few plants collected on the island is worth mentioning here. It is a golden-rod and seems to represent a species different from any heretofore collected. Interest centers in its relationships. Morphologically, it should apparently be most closely associated with *Solidago Boottii* of the southern foothills, Piedmont, and Coastal Plain, but which extends southward only to northern Florida. In habit, curiously enough, it most resembles a mountain golden-rod (*Solidago arguta*) of the Allegheny region and not known to grow south of North Carolina.

On Merritt's Island and the neighboring mainland shore are the southern outcrops of coquina. This rock, however, seems to have little effect on the vegetation, even though it occurs in sufficient quantities to afford a building stone, and is used quite extensively as such.

As we traveled southward, the more notable shell-middens were left behind, and the flora became noticeably different. While on the subject of shell-mounds we cannot refrain from remarking what an interesting subject for investigation their vegetation would furnish. These areas represent the latest developed native plant associations, for they were built up in recent times, certainly much later than any of the more recent geologic formations, of course excepting active dunes. Their plant covering developed during the process of building by the aborigines or later.

The composite family furnished the greatest number of species noted in flower. On the dry "scrub" the most conspicuous plant was a tall golden-aster (*Chrysopsis*) while in swamps and along streams a tick-seed (*Coreopsis*) eclipsed other flowering plants. The smallest woody plant of the sandy hills south of Eau Gallie was a scrub-oak of uncertain relationship. The shrubs were mostly about half a foot tall and in full fruit, the acorns being mostly about as numerous and as large as the leaves.

At Turkey Creek, near Malabar, we found a native *Portulaca* with tuberous roots, and an Asiatic vine, *Thunbergia fragrans*, that had escaped from cultivation and become naturalized. Along the estuary of Turkey Creek a lead-plant (*Amorpha croceolanata*) was in both flower and fruit. This shrub was described nearly a century ago, but overlooked until recently, when it was reinstated in our flora. Vegetation, however, was not plentiful there. The only other plant much in evidence was a sprawling form of the West-Indian prickly-pear (*Opuntia Dillenii*) which bore notably short fruits.

Midway down the peninsula, two odd plants, inhabitants of the "scrub," appeared and became progressively more abundant, both of them primarily winter bloomers. The one, a knotweed (*Thysanella robusta*), a low, rigid, somewhat woody herb with much-branched panicles bearing myriads of small flowers which heretofore we had found with only white flowers. This season, however, much of the bloom was a beautiful pink. The associate of the knotweed was a shrub-like mint (*Conradina*), and this was as graceful as the knotweed was stiff. It is irregularly branched and bears large, nodding flowers with peculiarly pink corollas mottled with magenta and of a very irregular pattern.

The grassy prairies and low pinewoods between the Sebastian River and the "scrub" were copiously adorned with hundreds and thousands of purple torches—the peculiarly purple inflorescence of a relative of the southern hound's-tongue or vanilla-plant which is closely related to the blazing-star. Its generic name is an anagram of the word *Liatris*, one of the botanical names of the blazing-star, and the plant in question is *Trilisa paniculata*. The only other species, *Trilisa odoratissima*, has been used as a substitute for vanilla, but it apparently possesses a flavor and odor more closely resembling that of the vanilla-substitute, the tonka bean, than it does that of the true vanilla, which is derived from the cured, unripe capsules of the climbing orchid, *Vanilla planifolia*.

This region and its vegetation was quite fully described in a former paper.[1] So suffice it to say here that the three most interesting plants found were a very small sedge (*Eleocharis*) in which the fruits remained attached to the spike-rachis after the scales had fallen away; a species of bushclover (*Lespedeza*) a most unlooked-for plant in that southern region; and the humus-loving orchid, *Habenella Garberi*, growing in the open sandy pine woods, instead of in the hammock, where alone the species had theretofore been observed in Florida.

In the heart of the old sand-dunes the predominating herb was the dog-fennel (*Eupatorium capillifolium*) It stood out in contrast with the other vegetation on account of its large masses of cluster-like plumes. The most prominent shrub was the so-called rosemary (*Ceratiola ericoides*). It was unusually attractive on account of the contrast between its deep-green and the snow-white sand in which it grew. This shrub seems to have two particularly fatal enemies; the one, fire, to which it succumbs quickly; the other, a parasitic vine (*Cassytha*) which is related to the laurel family in structure but in aspect and habit resembles dodder (*Cuscuta*). Scores of beautiful shrubs of the rosemary were buried under such a tangled mass of the parasitic vine that they were not only killed, but actually invisible, the skeleton of

---

[1] Journal of the New York Botanical Garden **20**: 205–207. 1919.

the shrub merely serving as a trellis for the vine which of course must soon also succumb for lack of nourishment.

In open places at West Palm Beach we found a mallow (*Sida cordifolia*), a plant new to the eastern coast. It has been known on the western coast for many years. Associated with it was the Barbadoes-gooseberry (*Pereskia Pereskia*), a climbing cactus armed with slender spines and bearing broad leaves. It is native in continental and insular tropical America and was discovered wild—natnralized—in Florida about two years ago.

The prairie-marshes along the way exhibited three plants that showed their inflorescences above the turf of grasses and sedges. They were: the purple *Trilisa*, already referred to, a yellow tickseed (*Coreopsis*), and the tall arrow-head, *Sagittaria lancifolia*.

Midway between Palm Beach and Miami there occurred the greatest surprise along the coast. In a flat part of the "scrub" what met our eyes but colonies of an Indian-pipe (*Monotropa*)! It is unlike the northern Indian-pipe (*Monotropa uniflora*) in being larger, and, further, it has more color. The base of the stem is pink, the middle part white, and the upper part, the upper leaves, and the floral parts, are cream-colored.

The afternoon of the day of our arrival in the Miami region was spent in the cactus garden of Mr. Charles Deering. All the plants under cultivation, except some of those from western American and Mexican deserts, were growing well, and the majority of the species were flourishing. One of the most interesting cases and one worthy of record is that of the giant-cactus (*Carnegiea gigantea*). Two plants of this species were introduced from the deserts of Arizona and set out in the garden last June. By the first part of December they had trebled in size. If this plant continues to thrive, as it has every appearance of doing, it may become one of the most conspicuous and interesting of the cultivated plants of southern Florida.

The following forenoon was devoted to the vicinity of Cutler, where some lichens, both epiphytic and epipetric (why not?), were collected. There, too, we saw a plantation of the saw-cabbage palm (*Paurotis Wrightii*), the first experiment of its kind, i. e., planted en masse, several months old.[1] This palm

---

[1] A single plant of the saw-cabbage palm has been growing in the nurseries of John Soar at Little River, Florida, for twenty odd years. Several isolated plants have been growing on the reservation of Charles Deering at Buena Vista for several years.

has the reputation of being difficult to transplant. However, although many of the main trunks of the individual plants seemed not to have survived the shock of transplanting, the stoloniferous branches at their bases were nearly all starting to grow. Several artificial groves of this palm which Mr. Deering is setting out at Cutler will doubtless soon form interesting plantations. If this attempt at cultivation proves successful, this beautiful palm will furnish a desirable element for decorative plantiug within the limits of its climatic endurance.

In the afternoon of the same day we proceeded to Royal Palm State Park, to which region we devoted two days in exploration.

The first early morning hours of our stay in Royal Palm Hammock were devoted to observations on birds. We arose several hours before sunrise and went eastward into the Everglades. First the night-feeding herons returned to their rookeries from their feeding grounds, then it was the turn of the day-feeding birds; for as the night-shift knocked off, these others left their rookeries and flew to feeding grounds of their own. The birds came and went on schedule time. The course of this daily migration is mainly up and down the sloughs forming the headwaters of Taylor River.

A day was spent in the Everglades between Royal Palm Hammock and West Lake, twenty-odd miles to the southwest of the hammock, or on the borderlands of the Cape Sable region.

A few miles southwest of the hammock are two very interesting phenomena. The first is botanical. It is the northern inland limit of distribution of the saw-cabbage palm (*Paurotis Wrightii*), already referred to. The outposts are scattered, comprising isolated colonies and colonies in Everglade hammocks. Further south, however, the palm is often the striking feature in the landscape. The second phenomenon is geological. It is more interesting than the first, yet it is invisible!

Among many interesting questions that arise in the course of our study of the plants in Florida, one in particular is often forced upon one's attention. It is this: was the flora in later geological times more extensive or more limited than it is at the present time?

We know that great changes have taken place in the plant-covering of the State. There is plenty of direct evidence that the white man has wrought destruction in the last few generations. We know that the Seminole Indian did his part in changing the vegetation through his mode of life and methods of hunting game. We are safe in assuming that the aborigines, perhaps a succession of aborigines, did their share in modifying the flora. If we do not have documentary evidence in the case of the aborigines, we do have circumstantial evidence in the shell-middens and other mounds.

The next question that arises, is what hand had nature in changing the flora? Has the land been elevated or depressed in relatively recent times?

Florida is well supplied with evidence of geologic changes, both superterranean and subterranean. Some of the rivers of the southeastern coastal region, for example, the New River and the Miami River, have deep channels that have been sculptured in the eastern rim of the peninsula. These would indicate that the land—at least on the eastern side—was formerly more elevated.

Now let us consider certain more recent and, as it seems to us, more interesting evidence.

During the last week of December 1917, I received the following information from Charles A. Mosier, custodian of Royal Palm State Park:

"Dredge on road is about four and a half or five miles from the park, making slow progress. It will be two years or more before it will reach Cape Sable. The dredge threw up some peculiar rock when they were working last week. One of the engineers called it petrified cypress trees, but it looks more like stalactites. Some of the bases must weigh a ton or more. I am anxious for you to come down as I wish to show them to you before the grading covers them."

The dredge referred to was one working on the extension of the Ingraham Highway which is planned to connect Miami with Cape Sable.

Southwest of Royal Palm Hamock the oolitic limestone in most places lies a few inches beneath the marl of the prairie. The water table, at its minimum, is relatively a few inches beneath the surface of the limestone. So, in making roads in that part of the Everglades, a dredge excavates the rock in front, according to the size of the road and depth of water desired. The shattered rock is thrown up on one side and a substantial road-bed and a navigable canal result from the one operation.

In the case of the above-quoted incident, a difference in the rock was noticeable in drilling preparatory to blasting. The blast shattered the top of a subaqueous cavern! Stalactites varying from the diameter of a finger to over four feet were thrown out. Unfortunately, but naturally, there were no perfect stalactites, as the shock of the dynamite had broken them into irregular sections. The dipper of the dredge, terminating a boom nearly thirty feet long, was let down into the cavern and swung around in all directions without encountering any obstructions. Here in the wet Everglades is a subaqueous cave. Yet the sections of the stalactites indicate great length and they could only have been formed in a cavern in which the floor, or at least the upper portion of the cavern, was elevated above the water table.

This is only one evidence of various hidden phenomena and structures in Florida—things submarine and things subterranean.

We are thus forced to believe that southern peninsular Florida was at no very remote date, geologically speaking, much more elevated than it now is, perhaps as much as fifty feet, possibly more. If this were the case, the present Florida reef on which the Florida Keys stand and the reef along the eastern coast out to the Gulf Stream, as well as, most likely, much of the present bed of the Gulf of Mexico adjacent to the western coast, was then dry land. However, just what the topography and the vegetation of southern Florida was at that time will remain a mystery forever.

All this brings up another question: Is the Florida peninsula at the present time rising or sinking?

The canal and road-bed of the new highway beyond Royal Palm Hammock are making accessible a region hitherto unvisited by naturalists.

The end of the canal at the time of our visit (December 1919) was in an arm of West Lake, which is a body of water of uncertain geographic position, on or near the Dade-Monroe County line, and not far from the Bay of Florida. The land is rather low there, but not very far to the westward there is more elevation. This constitutes the eastern extension of the Cape Sable region of Florida. Much of that region will be readily accessible within a short time and it will doubtless yield many interesting plants, as well as develop unusual problems.

In former times that region was one of the favorite hunting grounds of the Seminole Indians. It was a closed country to the white man, except to the more venturesome—particularly those who were in search of plume-birds. Even now if one happens to meet an Indian in the southern end of the Everglades and asks him where the plume-birds are he will invariably point in the direction of the Cape Sable region.

During our short stay in the low hammock about West Lake, we discovered an additional species of the wild-pepper (*Peperomia*) for the flora of Florida as well as for that of Continental United States, as the genus *Peperomia* does not occur in the States outside of Florida. Among other noteworthy finds were one of our rarer epiphytic orchids (*Oncidium sphacelatum*), growing in more luxuriant masses than we have seen it elsewhere in southern Florida, and the large epiphytic·cactus, a prickly-apple (*Harrisia Simpsonii*), which grew invariably on logs and tree-trunks, usually two to six feet above the hammock floor. Here, it had taken itself to the trees just as about Cuthbert Lake, where we found it several years ago, and where it sometimes occurred ten or twelve feet up in the mangroves.

Upon our return from the Cape Sable region we set out for the Lake Okeechobee region, whence we had hoped to continue across the peninsula to Fort Myers. Our course was the same as that described in a former paper. In brief, it was northward along the eastern coast to Jupiter, thence northwest through Hungry Land and the Alapattah Flats and the shore of Lake Okeechobee to Okeechobee City.

On the way we stopped long enough to visit the recently discovered Indian-pipe, and a further search in the vicinity of our first discovery brought to light a copious growth of that interesting plant in all stages of development, from mere buds pushing up through the sand to fully mature fruiting colonies with ripe capsules and dried stiff stems.

Striking westward into the wilderness at Jupiter and crossing some belts of "scrub," pineland, and streams bordered with hammocks, we came into Hungry Land with its numerous shallow ponds. Many of the ponds were inhabited by scraggy cypress trees which, however, were less stunted than those occurring southwest of Royal Palm Hammock referred to in previous papers.[1]

In Hungry Land the most conspicuous plant, as was the case on the prairies back of Sebastian, seventy-odd miles further north, was the false hound's-tongue (*Trilisa paniculata*). There were several other kinds of plants in bloom, but the region was not the showy flower-garden of the previous spring, as described in a former paper,[2] for, although the season was not winter in a technical sense, it was really the mid-winter of that region.

However, many plants were in evidence, although not ubiquitous as they would be in the spring. Among others was a closely prostrate kind of *Houstonia*, perhaps new to science. It bears small, roundish leaves and small, slender trumpet-shaped, white flowers.

About the cypress ponds were a tall pipewort with large, white, button-like head standing on top of the slender, ribbed stalk, also a bright-yellow flaveria, and some small asters, while the most elegant of all the southern asters, *Aster caroliniana*, clambered up into the cypress trees, and bore myriads of large showy heads of yellow and purple flowers.

On the edges of the ponds one may find more or less hammock growth, composed chiefly of swamp-bay (*Tamala*), Saint John's wort (*Hypericum*), gallberry (*Ilex*), queen's-root (*Stillingia*), and wax-berry (*Cerothamnus*), all of which were in flower, while in the ponds, the water-lily (*Castalia*) and the spatter-dock (*Nymphaea*) flourish.

Two plumose kinds of dog-fennel were in flower, the one, *Eupatorium capillifolium*, tall and coarse in the drier places, the other, *Eupatorium leptophyllum*, low and delicate, in the moister spots.

The mixture of rich soil and sand thrown up from the slough by the dredge along the Saint Lucie Canal supported a luxuriant growth of herbs and vines. Asters as high as one's head and dog-fennel (*Eupatorium capillifolium*) twelve feet tall attested to the fertility of the soil.

The Alappatah Flats on the northern side of the Hungry Land

[1] Journal of The New York Botanical Garden **21**: 53. 1920; **22**: 64. 1921.
[2] Journal of The New York Botanical Garden **22**: 30. 1921.

Slough were nearly flowerless. After passing some temporary Indian camps in the hammock near the canal we again came into the pinewoods. There, the two plants that were most visible in the failing light of the evening were the bleached skeletons of a composite (*Carphephorus*), which is related to the blazing-star (*Lacinaria*), and a wire-grass (*Aristida simpliciflora*), which grew in large colonies and was particularly in evidence on account of the persistent and tangled masses of the long awns of the inflorescence, which terminated the stems.

Before we had emerged from the flats the sun had set, and as a result of several unexpected previous and subsequent interruptions to our prearranged schedule, we reached Okeechobee City much nearer midnight than sundown. The chief causes of delay were a series of shallow sloughs on the border-land between the Alapattah Flats and the Okeechobee prairie, in the quicksands of which our cars bogged. Then, there was the necessity of making a new trail part way up the shore of Lake Okeechobee in place of the old trail which was then submerged on account of the partial filling up of the lake basin by the summer and fall rains.

The following morning brought us the information that the prairies between the Kissimmee River and the Caloosahatchee were flooded to such a depth that a detour of approximately one hundred miles would be necessary in order to get to Fort Myers. In fact, it was reported that there then was more surface water west of Lake Okeechobee than at any time in the past six years.

Consequently the expedition to the Fort Myers region was canceled. However, it was not the detour that decided us against carrying out our plans, as distance means but little nowadays in Florida; but the abnormally high water made the accomplishment of the main purposes of our proposed excursion impracticable. The deferred plans will be carried out at a future date when time is available and meteoric conditions are more favorable.

As an alternative we went to Fort Bassenger, which is situated northwest of Okeechobee City, by trail twenty-odd miles distant. The fort was located about a mile and a half south of the present crossing of the Kissimmee.

The present owner of the site where the old fort stood, a son of a soldier in the army during the Seminole Wars, who remained in that part of Florida, told us that when he settled there, about thirty years ago, relics in the form of parts of harness, horse-shoes, pieces of ammunition, and silver coins were quite abundant. By that time the stockade had disappeared, but a plow would often strike the stumps of old posts that still remained in the ground. It was at this point that Zachary Taylor crossed the Kissimmee River and on Christmas day, 1837, inflicted a severe defeat on the Indians, near Onothohatchee Creek during that unfortunate period of our domestic history. The latter stream now usually bears the name of Taylor's Creek, and is near the present site of Okeechobee City, which name supplants that of Tantie, the name of the earlier settlement there.

Back from the western bank of the river there is a sand ridge and a swamp. The ridge is covered with a scrubby growth of oak trees and miscellaneous shrubs. A half dozen kinds of oaks were collected and all of them were in fruit, a condition not often met with in these scrub-oak regions. Several mosses, characteristic of dry places, particularly a kind of *Campylopus*, were in abundance, as were also some lichens.

The swamp was overgrown with tall trees of the sweet-bay (*Magnolia virginiana*). The trees were so tall that from a distance the swamp had the appearance of a prominent elevation much higher than the sand-ridge, but it is really much lower. The sweet-bay trees showed a very dark-green when the air was still, but when disturbed by the breeze the whole bank of green was turned to a glistening silver-white on account of the pale under side of the leaves. Associated with the sweet-bay was the loblolly-bay (*Gordonia Lasianthus*). This kind of swampy hammocks are not unusual in the lower places in that region. In the flowering season they doubtless present an attractive sight on account of the numerous white flowers, those of the sweet-bay magnolia-like, those of the loblolly-bay camelia-like. There is often present, also, a holly (*Ilex Cassine*) with rich green leaves and bright red berries.

Returning to the bridge we recrossed the Kissimmee River, which at that season was full to overflowing. Not only was the channel filled, but the wide flood-plain was submerged. The flood-plain was, however, not deserted. The cattle that are accustomed to grazing on the turf that covers it in periods of dryer weather were now standing in the water feeding on the water-plants which formed great floating mats of succulent vegetation. These half-wild cattle are largely aquatic anyway. On several occasions in crossing the Kissimmee on former excursions, we have seen this Florida stock take a notion to cross to the other side of the river, plunge into a swift current, and swim to the opposite bank with more grace than they ever exhibit "on the hoof."

From the eastern side of the Kissimmee we headed for Okeechobee City. After traversing several miles of pineland in which there was little to attract the attention at that time of the year, except several yellow-blossomed members of the sun-flower family, we again came out on the prairies where few shrubs but numerous large and small herbaceous plants invited attention. The more prominent shrub, aside from the wax-myrtle, was a small-leaved holly, which undoubtedly represents a species not hitherto recognized.

Several low parts of the prairie, each many acres in extent, were covered with a pale grass with stems about a yard high and so closely set that the effect of a low-hanging mist was produced.

The smallest flowering plant on the low prairie was a member of the figwort family, with the generic designation of *Hemianthus* (meaning half a flower), referring to the very irregular and one-sided development of the corolla. It is readily visible only *en masse* when it forms bright-green, prostrate mats on the wet sand. It was in full flower where we found it, but the flowers are scarcely visible without the aid of a hand-lens. In strong contrast to this minute plant was a very large flag (*Iris*), the two plants often growing side by side. The flag, with leaves a yard tall or more, covers acres of prairie. The plants were approaching their flowering season, for we had found the large cucumber-like fruits in the same locality the previous spring. When in flower those vast areas of iris doubtless form seas of blue on account of the myriads of large blue flowers which are borne on stems three to four feet tall.

Four different plants of the sunflower family were in evidence. They represented rather closely associated genera: Two were fleabanes (*Erigeron quercifolius* and *E. vernus*). The third was an aster (*Aster Simmondsii*), and the fourth a boltonia (*Boltonia diffusa*). The aster and the boltonia resembled each other so closely in habit that it was not until the involucres and the fruits were examined that their relationships were evident.

The higher prairies were not without their herbaceous plants, flowering in season. One of the more conspicuous was a tall, slender goldenrod (*Solidago angustifolia*), with a greatly elongate inflorescence. It ranges from the seacoast to the interior. Another was a slender rayless-goldenrod (*Chondrophora nudata*) with a flat-topped inflorescence. In addition to these there was a large-headed composite (*Helianthella grandiflora*) resembling a sunflower.

The inconspicuous plants then in flower were an orchid relative—*Burmannia biflora*—with short, hair-like, minutely scaly stems mostly terminated with a pair of colored but very small flowers of peculiar structure, and a low-creeping spike-rush (*Eleocharis*).

About sundown, just as the moon-flowers and the morning-glories were opening for the night, we returned to Okeechobee City, spent the night there, and early the following morning set out towards the northeast for Fort Pierce.

Our first collecting ground was the hammock of Mosquito Creek, six miles east of the Onothohatchee or Taylor's Creek. There asters and goldenrods were in bloom, and the largest southern ladies'-tresses (*Ibidium cernuum*) was at the height of its flowering season. The larger plants were a yard tall and bore spikes of white flowers six to ten inches long.

The next stop was six miles further eastward in the hammock of Cypress Creek. This hammock is larger than that of Mosquito Creek. Both hammocks, however, have a copious growth of river-cypress (*Taxodium distichum*) in addition to the broad-leaved trees. There was more variety in the asters here, and a peculiar violet with white flowers appeared. Also, the ladies'-tresses mentioned above grew here in great luxuriance and also another terrestrial orchid (*Habenella Garberi*).

After a drive of six more miles we stopped to investigate a small low prairie where numerous but almost uniformly small-flowered plants grew intimately intermixed, all together forming a close turf. Many families from the grasses to the sunflowers were represented. Several interesting and critical species of nut-rush (*Scleria*) and of beak-rush (*Rynchospora*) were represented in abundance.

At least three kinds of bladderworts were there, but one, a member of the genus *Stomoisia*, was ubiquitous. It was a wiry-stemmed plant a few inches to two feet tall, with small yellow flowers. It is apparently a small-flowered form of *Stomoisia juncea*. The composite of those prairies, then in full bloom, was the common *Chaptalia semiflosculare* of the southern States, where it usually blooms in the spring. This locality is near to its southern geographic limit, for, a little further south the West Indian *Chaptalia dentata* takes its place in the flora. That small prairie harbored two species of ladies'-tresses,—*Ibidium Beckii* and *I. laciniatum*,—both, however, much smaller plants than the *Ibidium cernuum* mentioned above.

On these prairies the pond-cypress (*Taxodium ascendens*) replaces the river-cypress (*Taxodium distichum*). At several localities we learned from observation that the twigs produced from wounds on the lower parts of the trunks of the pond-cypress and on stumps bore foliage resembling that of the river-cypress. It may be of interest to record that a case similar to this happened several years ago on one of the trees of the pond-cypress cultivated in the New York Botanical Garden, where both species are hardy, although at a latitude considerably north of their natural geographic limits.

After a series of prairies were crossed we came to the pinelands and then to the "scrub" on some of the dunes of which Fort Pierce is built. Turning southward at Fort Pierce we reached Miami in the evening.

Next we visited the Deering Snapper Creek hammock and the Matheson Snapper Creek hammock for the purpose of collecting plants and studying the erosion of the limestone. Certain types of erosion in these hammocks correspond with those often met with in the pinelands and have an important bearing on some studies on the erosion of the Miami oolite which we will publish at a future date.

The Deering Snapper Creek hammock[1] was once one of the

[1] This hammock has recently become the property of Mr. Charles Deering and is being kept in its natural condition. The Deering hammock at Cutler, the monarch of the hammocks of the Everglades Keys, referred to on a subsequent page, is also being maintained by Mr. Deering in its natural state. Pumpkin Key, one of the unique islands of the reef, is also maintained in a state of nature by Mr. Deering.

Meccas of the Miami region, and is just as beautiful as formerly, so far as its vegetation is concerned. Furthermore, it contains the largest bustic-tree yet observed. The trunk-diamete- is about twenty-seven inches. There in the vicinity of this tree the first tropical filmy-fern (*Trichomanes punctata*) was found in the continental United States nearly twenty years ago.

Before the drainage fever became epidemic and rampant among the inhabitants of southern Florida, this hammock had other charms in addition to its vegetation. It possessed a superterranean stream and a subterranean stream, part of which, at least, emerged in a boiling spring. Both streams were part of the natural drainage of the Everglades. Now a drainage ditch has lowered the water table beneath the bed of the stream, and, it is said, by an error on the part of the engineer, the ditch was dug directly through the boiling spring. The stream above referred to represented the southern geographic limit for the quillwort (*Isoetes flaccida*), and until a recent date, the southern station for the hand-fern (*Cheiroglossa palmata*), which grew on the cabbage-trees (*Sabal Palmetto*) that stood along the stream. In the vicinity of the spring there once occurred the largest wild or semi-wild avocado trees (*Persea Persea*) and mango trees (*Mangifera indica*) in Florida, and also wild guava trees which bore nearly seedless fruits of a delicious flavor.

The Matheson hammock[1] has fully as beautiful a growth of shrubs and trees as the Deering hammock and it has not been directly devastated by a drainage ditch, although the lowering of the water table by a drainage canal in the adjacent prairie has changed the character of some of the shrubby vegetation, unfortunately not for the best. Moreover, there may be seen the remains of devastation wrought by a former generation. There the early settlers, like the recent homesteaders on the Everglade Keys, felled a large number of the giant trees of the hammocks merely to let them lie and decay through succeeding generations. It is remarkable how conscientiously and uniformly the pioneers wrought this devastation, seemingly just

[1] This hammock is being maintained in its natural condition by Mr. W. J. Matheson, who is also preserving Lignum Vitae Key from devastation. In this connection it should be mentioned that Mr. James Deering is preserving a part of the fast disappearing Brickell hammock from devastation.

as if it were one of the federal requirements for getting title to the land.

Each hammock of the Everglade Keys has one or more peculiar features in its vegetation. In the Matheson hammock there is an abundance of a very peculiar fern, a spleenwort (*Asplenium serratum*), with large simple leaves one to three feet long. This hammock represents the center of development of the fern. There is a more copious growth of it here than in all the other hammocks of the Everglade Keys put together and this locality doubtless represents the first place where the fern was discovered this side of the Gulf Stream, nearly half a century ago.

Fourteen miles of the barrier beach north of Miami were traversed for the purpose of comparing some of the dune plants with those of the Bahamian flora and for examining prickly-pears (*Opuntia*). There are numerous individually interesting kinds of flowering plants on the dunes between the ocean and Bay Biscayne comprising both rare and endemic species. The two most evident areas of plant associations were the wind-swept dunes on the ocean side and the mangrove hammock on the bay side. On the one hand the land was high and the shrubs and trees low, and on the other, land was low and the trees high. The most striking growth there is the mangrove hammock along the bay and the smaller lagoons. By a combination of the configuration of the land and the sweeping action of the winds, the mangroves are like a giant hedge. Starting low at the inner edge of the dunes, they slope gradually upward or rise in terraces just as if they had been trimmed by a pair of mighty shears. Among the mangroves are large areas of the leather-fern (*Acrostichum aureum*) and a copious growth of the rubber-vine (*Rhabdadenia biflora*), which climbs high in the trees and bears myriads of large white flowers which are most conspicuous in the forenoon. The rubber-vine, by the way, is properly so-called. It does yield rubber, but it is doubtful if this fact will ever make it attain commercial importance owing to the smallness of the vine. On the mangrove trees themselves grow several kinds of air-plants (*Tillandsia*) and our only endemic tree-orchid of a tropical type, *Encyclia tampense*.

Following these short excursions on the mainland our activities were transferred for three days to several of the keys of the

Florida Reef. We had the advantage of Mr. Hugh Matheson's fast motor-boat "Naisha" which Mr. Matheson navigated personally. A settlement named Islamorada on Upper Matecumbe Key was our first objective.

Once on Upper Matecumbe Key interesting plants were not long in coming to our notice. Two noteworthy sedges—galingales—were found in quick succession, the one, *Cyperus Pollardi*, new to the flora of the Florida Keys, and the other, *Cyperus Blodgettii*, hitherto known this side of the Florida Straits only on Key West, where it was originally discovered many years ago, but not since found in the United States. In recent years, however, it was discovered in the Bahamas and in Cuba.

During the first evening on the key we received a report of a planting of some "royal palms" that were found on Long Key, about a mile south of where we were stopping for the night. This report was both welcome and interesting, and our first move the following morning was to investigate the palms which proved to be as we had suspected, the hog-cabbage palm (*Pseudophoenix vinifera*), and not the royal-palm (*Roystonea regia*). This palm was found for the first time in Florida on Elliott's Key in 1886, and on Long Key some time afterward. It had been reported as extinct at both localities in Florida—a report since proved untrue; and rather recently it has also been found in the Bahamas on the keys of Cuba and in Hispaniola, where it was first discovered about the beginning of the eighteenth century.[1]

In the vicinity of the palms we found an abundance of a West-Indian pink-root (*Spigelia anthelmia*) which had been previously collected in Florida only on Elliott's Key, and specimens of the sweet alyssum (*Koniga maritima*), which so delights in the coasts of parts of Europe, the seed of which had probably been blown from nearby gardens during storms, and which had later sprung up in this truly maritime locality.

After settling the palm question we transferred our activities to Lignum-Vitae Key, which lies in Barnes Sound inside of and between the two Matecumbe Keys.

---

[1] A complete history of this palm will appear in a subsequent number of this Journal.

Lignum-Vitae Key was inhabited many years ago, as is evidenced by curious ruins, stone fences, and wells, and by exotic trees. One of the latter, a giant tamarind (*Tamarindus indicus*) measured about seven feet in circumference around the trunk five feet above the ground. The island is about to be improved and preserved by Mr. W. J. Matheson, who will protect all the original forest, which comprises many interesting tropical trees, some of them represented by good specimens.

Two conspicuous herbaceous plants were in bloom. On the exposed lower parts of the Key the sea-lavender (*Limonium brasiliense*) bore myriads of rose-purple flowers, while in the high hammock the native lead-wort (*Plumbago scandens*) sprawled over the rocks and shrubs well furnished with clusters of paler, but much larger flowers than those of the sea-lavender.

The Easter of the shrubs and trees had not yet appeared in the interior of the hammock, but around the edges the joe-wood (*Jacquinia keyensis*) was in full bloom, and the flowers filled the air with a delicious fragrance resembling that of the tropical buckthorn (*Bumelia angustifolia*), which, moreover, belongs to a family rather closely related to that to which the joe-wood is assigned.

Along the prairie-like areas, between the hammock and shore, the wild cotton trees (*Gossypium*), everblooming and everfruiting, and the dildoe (*Acanthocereus pentagonus*), then fruiting, together made tangled thickets. There we solved the origin of the specific name of the dildoe, namely, *pentagonus* i. e., five-angled. Almost invariably the mature stems of this cactus are stout and three-angled or three-sided, although the seedling plant starts with a slender, several ribbed shoot. The specimens on which the species was founded evidently had a five-angled stem, even if it is an exceptional condition. And there, on Lignum-Vitae Key, following many years of observations and search for such a specimen, we finally found one.

During the afternoon we returned to Upper Matecumbe Key and sought an area of original hammock for exploration. Although we found some rather tall forest, we are inclined to think it was second growth. Moreover, it is reported that the whole island some thirty years ago was cleared and planted with pineapples. Others of the Florida Keys were profitable pine-apple plantations in those days, before the lower eastern coast of the peninsula was readily accessible. The industry was later transferred to the mainland on the vast stationary sand dunes mainly in the vicinity of Saint Lucie Sound.

Upper Matecumbe Key is now largely planted with lime trees. However, we found a strange tree, unfortunately in leafage only, evidently a *Pisonia* relative. Also, we definitely established the occurrence of the tree cactus, *Cephalocereus Deeringii*, on that island by the finding of a single flower on one of the large specimens. This plant was discovered on Upper Matecumbe Key several years ago, but not until our visit was a flower obtained.

One of our endemic boneset relatives (*Osmia frustrata*) grew plentifully on both Upper Matecumbe Key and Lignum-Vitae Key. The plants were blooming and were conspicuous on account of the numerous blue or purplish heads which resembled little paint brushes. On the keys the plant is mostly only knee-high, but in the Cape Sable region, on the mainland, it often grows up to six or eight feet.

It was unanimously decided to devote the forenoon of our last day on the Keys to an examination of Indian Key. This historic island is on the reef outside of the main line of keys and opposite the interval separating Upper Matecumbe and Lower Matecumbe Keys. It is nearly circular in outline and comprises about six acres. It was once a port-of-entry. The island became notable through the activities and the subsequent death of Henry Perrine,[1] who settled there in 1838, making it both a

[1] Henry Perrine was born April 5, 1797, at New Brunswick, New Jersey. In youth he taught a school at Rocky Hill, N. J., then studied medicine at Philadelphia. After five years (1819–24) as a physician at Ripley, Illinois, and three (1824–27) at Natchez, Mississippi, he spent ten years (1827–37) as United States Consul at Campeche, Yucatan. While there he devoted himself with great singleness of purpose to securing for transportation to the United States various plants of economic value. This was peculiarly difficult, because the local officials were very much opposed to the export of living plants of economic value, and it was only by Perrine's influence as the most popular physician in the region that he was able to overcome the obstacles placed in his way. Upon his return to the United States, in 1837, he spent six months at Washington pressing his claims for Congressional assistance in his schemes for the introduction and propagation of tropical plants in the United States. In 1838, he received a grant of a township of land on Biscayne Bay, Florida, but the disturbances occasioned by the Seminole War,

PLATE 265

In Deering hammock, Cutler. Natural rock pedestal and flower-pot in foreground; jungle of cocoaplum (*Chrysobalanus*) in background. The one-time uniform layer of oolitic limestone, represented, approximately, by the top of the rock-structure has been eroded and removed mainly by the seasonal fluctuations of the water-table. The pedestal represents a hard core of rock that has resisted the processes of erosion. The surface is a rock filigree, the structure of which is more plainly shown in the insert. Ferns, in this case the sword-fern, herbaceous and woody flowering plants occupy the tops of pedestals and tables.

temporary residence and a nursery, pending the end of the Seminole Wars, when he contemplated moving to a large grant of land on the Florida mainland. Before he could consummate his long prepared and cherished plans, he was murdered by a band of intoxicated Indians who were being crowded southward by the advance of the white man's civilization.

Only three things remain on the key to indicate the improvements of a century ago. They are the masonary foundations of the former buildings, some commonly cultivated tropical trees, and, what is much more interesting, numerous descendants of the sisal (*Agave rigida*) plants Dr. Perrine evidently intro- ·duced as part of his nursery stock.

The stone slabs once placed near the middle of the key to mark the graves of Perrine and others, intact until quite recent years, have of late been destroyed or removed, perhaps, by vandals or treasure hunters.

Several modern frame houses, now deserted and not only unprotected, but plundered of their contents, ready to be consumed with the first fire that sweeps the island, stand on the higher part of the key.

We returned to Upper Matecumbe Key at noon. Our plan to stop on Tea-table Key, which lies near the outer side of Upper Matecumbe, and was at one time a naval base, was defeated by lack of time. We reached Miami in the late afternoon, the "Naisha" making the record run of eighty-odd miles in four hours and twelve minutes.

The day following our return from the Keys was devoted to· collecting flowerless plants in the Deering hammock at Cutler. The moister parts of this hammock yielded numerous species of lichens, hepatics, and mosses.

The erosion of the limestone in the lower parts of this hammock is different from that of any of the other hammocks of the Everglade Keys. The rock is a very pure limestone, with scarcely any sand in its composition. In ordinary rainy sea-

then in progress, prevented him from occupying his grant. He did settle however, upon the neighboring island, Indian Key, where he was killed by a band of marauding natives, August 7, 1840. After killing him, the Indians burned his house, destroying all his manuscripts and collections. His family escaped, and he might have done so if he had not hoped to dissuade the Indians from doing harm.—John Hendley Barnhart.

sons the water-table rises, and water fills the area to a depth of two or three feet. The water is essentially stationary sometimes for long periods. Whatever movement there is, is not horizontal, but perpendicular, and, of course, very slow, but it is this perpendicular movement of the water that causes the curious, often fantastic, results of erosion. Naturally, the maximum duration of submergence is near the floor of the hammock, so that the water charged with the acids of decaying vegetable matter and with carbon dioxide works longer and stronger on the lower parts of the rock than on the upper. As a consequence; the maximum erosion is near the hammock floor and we find urns, flower-pots, tables, pulpits, cañons, caves, tunnels, natural bridges, all of which, together with the accompanying arboreous vegetation, particularly the buttressed pond-apple (*Annona*) trees, constitute a series of grottoes different from any others in the State.

The uneven structure of the rock, hard and soft, results in an uneven surface resembling lace-work or filigree. What loose material is dissolved out of the limestone must of necessity fall to the hammock floor, for there is no current of water or other means to transport it. The hammock floor is usually nearly or quite free of sand, which would naturally accumulate there if it came out of the rock. The floor is mostly covered with humus formed from the continuously decaying vegetable matter.

Upon returning to Miami that evening, it was decided to start north along the eastern coast for a two days' excursion, particularly for further studies in cacti and wild-pepper plants, for a rapid survey of the vegetation along the way, for photography, and to prosecute some scout work for future investigations.

We traveled as far as Stuart before midnight and ran into a "norther." Being uncomfortably cool we stopped there and helping ourselves to some unoccupied rooms in the principal hotel we spent the rest of the night comfortably.

Early in the morning we set out for the hammock along Saint Lucie Sound in order to get a photograph of a wild-pepper plant (*Peperomia humilis*); however, the strong gale of the "norther" which was against us decided us to drive on to Daytona where we arrived just before sunset.

All the way from Fort Pierce to Daytona the effect of the chill of the "norther" could be seen on the flowers of the moon-vine. During the previous night we had noticed that the moonflowers along the way had not opened. Early in the morning the buds had for the most part failed to open; but as the sun warmed the atmosphere the buds burst forth and we had moonflowers all day long, as a result of the retarding effect of the low temperature of the preceding night. Thus one also may have a night-blooming cactus flower in the day-time if the bud be put on ice during the night in which it would normally open.

Notwithstanding the fact that winter conditions prevailed generally, technically or according to the almanac winter had not come; it was really the last weeks of autumn. However, along the way, particularly in the vicinity of Vero, the cinnamon-fern (*Osmunda cinnamomea*) had skipped winter and indicated spring. The tall red spore-bearing leaves were in their glory, and in addition, a score of spring-flowering herbs were associated with it. The spruce-pine (*Pinus clausa*) had come into flower during the two weeks' interval since we had passed through the same country, and its yellow flower-cones were in prominent clusters among the leaves.

A scattering of tropical shrubs was observed along the lagoons, particularly those with baccate fruits. They are probably sown by migratory birds, and finding congenial localities, sprout and survive. However, they never grow with the vigor they exhibit near the southern end of the peninsula. Among those in flower at that time was the marlberry (*Icacorea*) and the myrsine (*Rapanea*). The former shrub occurred as far north as Daytona.

On this ride two things conspired to make one think he was in a higher latitude than Florida. First, there was the cold stiff breeze of the "norther." Then, the fruiting plants of the tropical hemp-vine (*Mikania cordifolia*) suggested, especially when one was half numb as a result of riding against the cold wind, the fruiting vines of the virgin-bower (*Clematis virginiana*) of the North.

One quite unusual sight presented itself at several places on the old dunes which formerly had been under cultivation as pineapple fields. A fern had become a rampant weed. Many acres of former fields had been taken possession of by the brake

(*Pteris caudata*) which grew there to the exclusion of nearly all other vegetation. The rapidly growing underground stems of this fern enable it to spread rapidly in the loose sand and to crowd out most of the plants that would naturally grow with it.

Returning south again we were able to complete the record of a section of the vegetation on the mainland facing the various lagoons along or near the Dixie Highway from Daytona to Miami or from the Halifax River to Bay Biscayne.

From Daytona to New Smyrna the Halifax River is lined with hammock, usually close to the water's edge. From New Smyrna to Oak Hill hammocks predominate, but some pineland is interposed. About sixteen miles south of New Smyrna, near Oak hill, wide marshes with palmetto hammocks not only occur near the lagoon, but also extend far inland; thence to Titusville hammocks and stretches of pineland appear again.

For the distance of twelve and a half miles south of Titusville are stretches of "scrub" and pineland interrupted here and there by small areas of hammock. Between five and six miles north of Cocoa, hammock appears and extends southward about fifteen miles or to about eight miles south of Cocoa. Thence there is "scrub" or pineland to Eau Gallie, where there are small areas of hammock, particularly about the mouth of Elbow Creek. South of Eau Gallie there is "scrub" and pineland to near Melbourne, where there is a little hammock about the mouth of Crane Creek. Thence southward we find stretches of "scrub" and small hammocks to the hammocks at the mouth of Turkey Creek. South of Turkey Creek as far as Malabar there is "scrub" and low oak hammock. Thence southward alternating pineland and "scrub" with a hammock about six and a half miles south of Melbourne.

Below Grant there is some hammock, then about a mile of "scrub" and some hammock near the fifteenth mile post south of Melbourne. Again, some "scrub" and then the hammocks about the mouth of the Sebastian River about eighteen miles south of Melbourne. One mile south of the Sebastian River, there is a small area of "scrub," thence pineland and hammocks with palmettos to Quay. Thence "scrub" or "scrub" and pineland mixed, and an occasional piece of hammock in low places to Vero. South of Vero "scrub" or near Oslo some pineland and hammock reaches the lagoon. "Scrub" is in evidence or predominates thence to Fort Pierce.

Remains of aboriginal shell-midden on Halifax River. There the red-man once feasted. Accumulations of shells, mainly oyster, clam, and conch, representing the habitations or rendezvous of aborigines, often occur along the coasts of Florida. These prehistoric monuments vary in size from a mere handful of shells to mountain-like masses up to eighty feet high. They are covered with an arboreous growth, and devoid of soil; the only matter aside from the shells is disintegrated shell material, charcoal, bones of animals, and the humus derived from the decay of the vegetation that has clothed them. Unfortunately the mounds are being drawn on for road material. With their destruction, interesting plant associations disappear.

South of Fort Pierce hammock appears again and extends along the shores of Saint Lucie Sound for a distance of about sixteen miles. Some of this hammock is apparently in its primeval condition, but much of it has been partly or wholly cleared. From the vicinity of Jensen and Rio, "scrub" on high rolling sand-dunes, with here and there a little hammock growth, extends to the estuary of the Saint Lucie River, where, of course, there is hammock. From the Saint Lucie River (Stuart) southward for about seven miles, there is a succession of pinelands and "scrub," with the two pines—*Pinus clausa* and *P. caribaea*—more or less mixed. From about twenty-nine miles north of West Palm Beach to the twenty-first mile post, there is a long stretch of nearly or quite pure "scrub," then some pine-woods and "scrub" to the Jupiter River. Pine woods appear on the southern side of the river for a short distance, then "scrub" to about nine miles north of West Palm Beach, where there is a small area of pineland. After that "scrub" appears again and extends all the way to West Palm Beach. South of West Palm Beach there is "scrub" with patches of *Pinus caribaea* for about twelve miles and then pineland to two miles south of Boynton. After the pineland "scrub" extends to a short distance south of Pompono, where there are some hammocks and cabbage-trees and cypress swamps. Then the prairies, hammocks and cypress strands in the vicinity of Cypress Creek four or five miles north of Fort Lauderdale, pass into a similar country about the forks of Middle Creek where there is also some "scrub" in the pineland. Thence low pinelands with some cypress extend to Fort Lauderdale. The hammocks of New River give way to pinelands and these, about two miles south of the town, are bordered with wide marshes. At a point about twenty-two miles north of Miami is hammock, then pineland appears, but soon gives way to "scrub," which is interrupted with pineland, but at eighteen and a half miles it appears again and extends to near Hallandale, sometimes with a pure growth of *Pinus clausa*, at other times with *Pinus clausa* and *P. caribaea* growing together. From Hallandale to Miami there is mostly pineland, except along the intersecting creeks, which naturally are bordered by hammock.

The time of our visit to Florida was naturally the dullest

season of the year, as far as flowering plants were concerned. It was just between the end of the fall flowering season and the resurrection characteristic of spring. Of course, there were herbaceous plants in bloom nearly everywhere, either individually or in certain areas more than in others. However, nowhere were there shows of flowers as there are at other seasons of the year.

The flora of the Everglades is at the best limited and never conspicuously showy. That of the pinelands in spring and summer is copious and usually very brilliant. However, there is one general locality where the collector will find many plants in flower at any season of the year. It is the borderland where the pinelands and the Everglades meet. There one is sure to find plants to interest him. So, at a prairie-like area along the Tamiami Trail west of Miami on our last collecting excursion, we found a garden with scores of plants in full bloom. The kinds ranged from grasses to composites. Most of them were low, ranging from pipeworts on the one hand, an inch or two tall, to, on the other, an Indian-plantain, between two and three feet tall. The latter was the most striking plant in the collection, and with its glaucous, elongate, sharply toothed leaves and its flat-topped inflorescence of many pale heads, and also its large numbers, it stood out in strong contrast with all the accompanying vegetation.

Many genera were represented by a single species, some by more than one species. Several of the more conspicuous plants fell into genera by pairs; for example, there were two kinds of panic-grass (*Panicum*), two galingales (*Cyperus*), two spike-rushes (*Eleocharis*), two nut-rushes (*Rynchospora*), two blazing-stars (*Lacinaria*), two flea-banes (*Erigeron*), two goldenrods (*Solidago*), and two marsh-fleabanes (*Pluchea*).

Much of interest has been lost from the historic trails of Florida. An infinite interest yet remains.

JOHN K. SMALL

THE PALM COLLECTION

During and after the recent reconstruction of the roof of the large palm greenhouse at Conservatory Range No. 1, New York Botanical Garden, the extensive collection of palms was rearranged. This collection has been brought together during the past twenty years, partly by gifts of fine plants from many friends of the institution, partly by specimens obtained in exchange with other gardens and partly by plants obtained by exploring expeditions and partly by plants grown from seeds; it now contains about 130 different species, represented by 530 individual plants, large and small and, except for a few small specimens in the propagating houses it is installed in houses 1, 13, 14, and 15 of Conservatory Range No. 1, near the Bronx Park Station of the Third Avenue Railroad. There are many noteworthy and perfect specimens.

The two tallest specimens are a feathery Cocos from Brazil, reaching a height of forty feet and a Corozo Palm (*Acrocomia*) from Porto Rico, its trunks a foot in diameter, entirely covered with black, sharp spines.

The most graceful one under the large dome is a rock date palm, each of its ten-foot leaves arching nearly to the ground. Two sturdy companions of this are a pair of *Phoenix reclinatus*, spreading their leaves for thirty feet or more from thick old trunks, which are now nearly three feet in diameter. A sugar palm of India, *Arenga saccherifera*, which is of economic importance there, will also be found under the big dome. Starting up the sides of this huge glass house are now many young climbing palms of the genus *Nunezharia*. These will be trained upwards. In the houses next the dome will be found numerous smaller specimens, many of them rare. Here may be seen the Wax Palm from Brazil, with waxy white under-surfaces of leaves; this palm will be one of the most striking in the collection in future years.

In House 13, the eastern secondary dome, are two notable groups. One is of our American Desert Palm, *Neo-Washingtonia robusta*, four magnificent plants from the cañons of Southern California; another of the Chinese Fan Palm, the leaves measuring five feet across. In this house also will be found the tallest palm of Porto Rico, the Plume Palm, which was brought to the Garden in 1906 by Dr. Britton and the late Professor J. F. Cowell.

CONFERENCE NOTES FOR NOVEMBER AND DECEMBER

The November Conference of the Scientific Staff and Registered Students of the Garden was held in the museum building on the afternoon of November 2nd, 1921.

Two topics were represented, one by Dr. Arthur Hollick and one by Dr. F. J. Seaver, who have prepared the following abstracts of their discussion.

"A REVIEW OF THE FOSSIL FLORA OF THE WEST INDIES"

BY DR. ARTHUR HOLLICK

Very little is known about the fossil flora of the West Indies. The complete bibliography of the subject consists of less than twenty titles. Some of these are mere incidental references to the occurrence of fossil plants, without descriptions or names; others are descriptions of silicified wood, of doubtful diagnostic value; others are descriptions of marine calcareous algae contained in limestones.

Antigua and Cuba have furnished most of the woody specimens; references to leaves and other remains of plants are to be found mostly in connection with descriptions of Santo Domingo and Trinidad; the algae are from Antigua, Anguilla, Saint Bartholomew, and Martinique. I am aware of only one article of a descriptive nature, with illustrations, based upon identifications of fossil leaves.[1]

In the Mueseum of the New York Botanical Garden we have collections of well defined, identifiable leaves from three localities in Cuba, from three in Porto Rico, from three in Trinidad, and from one in Santo Domingo. Those from Cuba, Porto Rico, and Trinidad were all collected within the past six years. The collection from Santo Domingo, consisting of seven specimens only, was made by Wm. M. Gabb, in 1868, and was only recently brought to light while I was engaged in a search for all available material representing the West Indian fossil flora. The Trinidad specimens were collected, in part by Dr. N. L. Britton in 1920, and in part by Gilbert Van Ingen in 1921. The

[1] Berry, E. W.: Tertiary Fossil Plants from the Dominican Republic. U. S. Nat. Mus., Proc. 59: 117–127, *pl. 21.* 1921.

Porto Rico specimens were collected by Bela Hubbard in 1915, incidental to the geological survey of the island under the joint auspices of the New York Academy of Sciences, the American Museum of Natural History, and the Insular Government of Porto Rico. The Cuban specimens were collected by Brothers Leon and Roca in 1918.

The total number of specimens in all the collections is about 200, of which about 50 are identifiable species or genera. Most of these have been figured and a number of provisional identifications have already been made. It is hoped that all identifications and descriptions will be completed in the near future.

The geologic age of all the collections is apparently Tertiary. The Porto Rican specimens are probably Eocene. The Trinidad and Santo Domingo specimens appear to be more recent— probably Miocene. The Cuban specimens are the most recent and may represent, at least in part, merely remains of the vegetation now in existence there, incrusted with calcareous tufa.

"PRELIMINARY NOTES ON TRINIDAD FUNGI"

By Dr. F. J. SEAVER

During the six weeks spent in Trinidad, six hundred and forty seven collections of fungi were obtained. In addition to these a small collection of slime moulds was made which collections were unnumbered, not knowing in what condition they would arrive. When divided up into exchange sets there will be more than two thousand specimens.

Trinidad is very rich in fungi in general, but very poor in certain groups such as the fleshy cup fungi and agarics. The cup fungi seemed not to be abundant even when the conditions of moisture and substratum were apparently most favorable. Whether this is a seasonal or general condition it is impossible to know from a single visit. A few large cup fungi were, however, obtained, among them *Peziza badia* which was found in abundance in an abandoned pit where clay is burned to be used for surfacing the roads.

The Pyrenomycetes, on the other hand, are abundant and many of them very different from the forms which occur in the North. Of the Hypocreales, a number of species of *Hypocrella* were collected. This genus is of interest for two reasons: In the first place, it represents the perfect stage of the genus *Aschersonia* and several papers have recently been published on this phase of the question. In the next place both *Hypocrella* and *Aschersonia* are of interest because they occur as parasites on destructive insects, especially scale insects. Such fungi are of importance because of the possibility of using them as a means of controlling the insects on which they occur. A few insects are held in check by this means. A number of species of Hypocreales were collected which have not been determined and may be undescribed. Others are South American species not before represented in our collections.

A large number of wood fungi were obtained most of which have been determined by Dr. W. A. Murrill. A few are apparently South American, not occurring in the North and can be determined only by comparison with South American material.

Considerable attention was given to the collection of the rusts on account of the unusual opportunity of getting the hosts determined in the field by Dr. Britton, the determination of the hosts being one of the prerequisites for the successful naming of the species; also for the reason that no attempt has ever been made to enumerate the species occurring in Trinidad. One hundred and sixty collections were made, representing seventy one species, four of which are new, and several others rare. A list of the species will be published by Dr. J. C. Arthur in the January issue of Mycologia. At least five sets of the rusts will be made, one hundred and fifty of which have already been exchanged for an equal number of tropical rusts collected by Dr. E. W. D. Holway.

This report is entirely preliminary since many of the collections of ascomycetes have not yet been studied critically.

The December conference of the Scientific Staff and registered students of the Garden was held on the afternoon of December 7th.

Mrs. E. G. Britton gave an interesting discussion of her extended studies of the mosses of Cuba, Haiti, and Trinidad. Numerous specimens were exhibited, particularly of the genera of pleurocarpous mosses.

Among them were included type specimens of several new

species, including *Daltonia* and *Stenodictyon*. Further notes on the species of *Rhacopilopsis* from Trinidad, French Guiana and Africa were shown and it was stated that the variations were such as to scarcely be specifically distinct.

Dr. N. L. Britton exhibited a set of ferns and flowering plants collected by the distinguished Swedish botanist, Dr. Carl Skottsberg on the islands of Juan Fernandez, Mas-a-tierra and Mas-a-fuera. Dr. Skottsberg has now completed a survey of the vegetation, which he finds consists of about fifty species of Ferns and Fern-allies and a somewhat larger number of flowering plants. Dr. Britton made special mention of the genus *Robinsonia* which includes trees and shrubs of the Compositae. In the naming of the species from these islands the names Robinson and Crusoe are frequently used in commemoration of the ship-wrecked sailor of Defoe's well-known narrative. The isolation of the islands together with their volcanic origin give special interest to the flora, which includes many species growing there only.

Dr. Skottsberg has also visited Easter island, which lies still farther to the West, for the purpose of special study of its vegetation.

A. B. STOUT.
Secretary of the Conference.

NOTES, NEWS AND COMMENT

Greatly needed repairs were made this fall on the central dome of Conservatory Range No. 1 and were successfully completed just before the cold weather started in. After some twenty years of exposure to the warm, moist atmosphere beneath, it was found necessary to replace all the wooden rafters and bars in which the glass is set and repaint the iron work of the entire lower dome. Much broken glass also had to be replaced. The work required the labors of some 12 or 15 men for a period of about seven weeks.

The commencement has been made this autumn in bringing together an extensive collection of Paeonias in the Horticultural Garden on land west of the principal plantations of Iris. Noteworthy gifts of plants for this purpose have been received from

Mrs. Edward Harding, being roots of fifty choice, recently developed varieties, which have been planted in a plot by themselves, and also seventy-five plants contributed by Mrs. Charles D. Dickey, planted in another plot. Mrs. Harding, who is a high authority on Paeonias, has given us valued advice as to location, soil and fertilizers, and through her interest the bonemeal fertilizer required was given by The American Agricultural Chemical Company.

About 500 biology pupils of Evander Childs High School visited the Garden with their teachers on December 6 and 7. Mr. Hastings gave a lecture on forestry, and considerable attention was devoted to the museum and greenhouse collections under the guidance of members of the staff.

Dr. L. O. Overholts, of the Pennsylvania State College, spent the latter half of December at the Garden, completing his study of *Pholiota*, an important genus of the fleshy fungi, for early publication in North American Flora.

The following visiting scientists have registered in the library during the autumn: Mr. D. S. Carpenter, Middletown Springs, Vt.; Miss Eloise Gerry, Madison, Wis.; Mr. L. J. Pessin, Agricultural College, Miss.; Prof. J. B. S. Norton, College Park, Md.; Prof. C. C. Glover, Ann Arbor, Mich.; Mr. John M. Arthur, Yonkers, N. Y.; Prof. W. W. Rowlee, Ithaca, N. Y.; Rev. J. P. Otis, Marshallton, Del.; Prof. Irving W. Bailey, Boston, Mass.; Dr. Ralph E. Cleland, Baltimore, Md.; Dr. Edgar T. Wherry, Mr. W. W. Eggleston, Dr. J. N. Rose, Mr. John W. Roberts, Mr. G. Hamilton Martin, Jr., Dr. Perley Spaulding, and Mr. Maurice Ricker, Washington, D. C.; Mr. A. A. Pearson, Treas. British Mycological Society; Mlle. Sophie Satin, Dresden, Germany, and Prof. N. T. Vavilov, Petrograd, Russia.

Meteorology for November: The total precipitation for the month was 4.60 inches. There were traces of snow on the 14th. The maximum temperatures recorded for each week were as follows: 63° on the 1st, 64° on the 7th, *70° on the 18th*, and 58° on the 22nd. The minimum temperatures were: 30° on the 6th, *28° on the 13th*, 32° on the 16th and *28° on the 26th*.

Meteorology for December: The total precipitation for the

month was 3.19 inches, of which 0.85 inches (8 and ½ inches by snow measurement) fell as snow. The maximum temperatures recorded for each week were as follows: *60° on the 1st*, 47° on the 10th and the 11th, 54° on the 18th, 45° on the 19th and 43° on the 27th. The minimum temperatures were: 23° on the 5th, 18° on the 9th, 13° on the 16th, *6° on the 22nd* and 7° on the 30th. The first ice of the autumn formed across the middle lake in the night following the 4th.

Meteorology for the year 1921: The total precipitation for the year at the New York Botanical Garden was 34.90 inches. This was distributed by months as follows: January, 2.39 inches (including 2 inches snow measurement): February, 3.23 (including 16 inches snow measurement); March, 2.22; April, 3.21; May, 2.62; June 3.02; July, 1.76; August, 4.73; September, 3.35; October, 0.58; November, 4.60 (including a trace of snow); December 3.19 (including 8 and ½ inches snow measurement). The total fall of snow for the year was 26 and ½ inches. The maximum temperature for the year was 98°, on the 22nd of June. The minimum was 4°, on the 19th of January. The first killing frost of the autumn was on the morning of the 26th of October, when a temperature of 29½° was recorded. The latest freezing temperature of the spring was on the morning of the 12th of April, when a temperature of 30° was recorded.

ACCESSIONS

84 specimens of fossil plants from Trinidad, West Indies. (Collected by Dr. N. L. Britton and Professor Gilbert van Ingen.)

34 specimens of flowering plants from Texas. (By exchange with Professor Albert Ruth.)

347 specimens of ferns from Haiti. (By exchange with the United States National Museum.)

150 specimens of fossil plants from Brazil. (By exchange with Professor J. C. Branner.)

1 photograph of the type specimen of *Pentstemon mensarum*. (By exchange with the United States National Museum.)

12 photographs of specimens of cactaceae. (Acquired from Mr. N. E. Brown.)

63 specimens of mosses from Haiti. (By exchange with the United States National Museum.)

6 specimens of *Rubus, Viburnum, Tilia*, and *Amelanchier* from the southern United States. (By exchange with Mr. W. W. Ashe.)

88 specimens of ferns from various localities. (By exchange with the United States National Museum.)

1 specimen of *Polygonum neglectum* from Staten Island, New York. (Given by Dr. N. L. Britton.)

22 specimens of cacti (2 photographs). (By exchange with the United States National Museum.)

INDEX

Provisions for
Benefactors, Patrons, Fellows, Fellowship Members, Sustaining Members, Annual Members and Life Members

---

1. Benefactors

The contribution of $25,000.00 or more to the funds of the Garden by gift or by bequest shall entitle the contributor to be a benefactor of the Garden.

2. Patrons

The contribution of $5000.00 or more to the funds of the Garden by gift or by bequest shall entitle the contributor to be a patron of the Garden.

3. Fellows for Life

The contribution of $1000.00 or more to the funds of the Garden at any one time shall entitle the contributor to be a fellow for life of the Garden.

4. Fellowship Members

Fellowship members pay $100.00 or more annually and become fellows for life when their payments aggregate $1000.00.

5. Sustaining Members

Sustaining members´ pay from $25.00 to $100.00 annually and become fellows for life when their payments aggregate $1000.00.

6. Annual Members

Annual members pay an annual fee of $10.00.
All members are entitled to the following privileges:

1. Tickets to all lectures given under the auspices of the Board of Managers.
2. Invitations to all exhibitions given under the auspices of the Board of Managers.
3. A copy of all handbooks published by the Garden.
4. A copy of all annual reports and Bulletins.
5. A copy of the monthly Journal.
6. Privileges of the Board Room.

7. Life Members

Annual members may become Life Members by the payment of a fee of $250.00.

Information

Members are invited to ask any questions they desire to have answered on botanical or horticultural subjects. Docents will accompany any members through the grounds and buildings any week day, leaving Museum Building at 3 o'clock.

Form of Bequest

I hereby bequeath to the New York Botanical Garden incorporated under the Laws of New York, Chapter 285 of 1891, the sum of.........

CPSIA information can be obtained
at www.ICGtesting.com
Printed in the USA
BVHW04*1058170918
527708BV00014B/1385/P